CAMBRIDGE COUNTY GEOGRAPHIES

General Editor: F. H. H. GUILLEMARD, M.A., M.D.

NORTH LANCASHIRE

Cambridge County Geographies

NORTH LANCASHIRE

by

J. E. MARR, Sc.D., F.R.S.

With Maps, Diagrams and Illustrations

Cambridge :

at the University Press

1912

CAMBRIDGE UNIVERSITY PRESS
Cambridge, New York, Melbourne, Madrid, Cape Town,
Singapore, São Paulo, Delhi, Mexico City

Cambridge University Press
The Edinburgh Building, Cambridge CB2 8RU, UK

Published in the United States of America by Cambridge University Press, New York

www.cambridge.org
Information on this title: www.cambridge.org/9781107688520

© Cambridge University Press 1912

First published 1912
First paperback edition 2013

A catalogue record for this publication is available from the British Library

ISBN 978-1-107-68852-0 Paperback

PREFACE

I WISH to thank Mr K. J. J. Mackenzie, M.A., for information concerning the agriculture of North Lancashire.

I am, as ever, grateful for the courtesy of all connected with the University Press with whom I have been brought into contact while the work was in progress, and must specially thank Mr H. A. Parsons who undertook the production of the diagrams at the end of the book.

J. E. M.

May, 1912.

CONTENTS

ILLUSTRATIONS

The illustrations on pp. 17, 41, 59, 64, 136, 140, 153 are
from photographs by Mr Herbert Bell of Ambleside; those on
pp. 7, 15, 26, 30, 31, 51, 57, 60, 96, 98, 100, 104, 106, 115, 128
by Mr Brunskill of Bowness-on-Windermere; those on pp. 2, 9,
12, 20, 28, 35, 40, 43, 45, 47, 53, 54, 55, 66, 94, 101, 122,
130, 133, 135, 138, 142, 145, 151, 161, 162, 164, 165, 167, 168,

169, 170, 171, 173, 174, 176 are from the series of Messrs F. Frith & Co., of Reigate; the photographs on pp. 81 and 126 were kindly taken for me by Mr W. M. Rankin, Principal of the Storey Institute, Lancaster; the illustrations on pp. 118, 123, 125 are from photographs by Mr W. Sumner of Lancaster.

I am indebted to Mr W. Ralph Peel of Knowlmere Manor, Clitheroe, for the illustration on p. 92, and to Mr C. C. Madeley, Director of the Municipal Museum, Warrington, for that on p. 114; the photograph on p. 72 was taken by Mr A. Horner of Settle, to illustrate a work by Mr Reginald Farrer, and I am indebted to Mr Farrer and Mr Edward Arnold for the right of reproducing it. The portraits on pp. 155, 157 and 158 are from photographs by Mr Emery Walker. The photograph on p. 119 was taken by Mr Edwin Wilson of Cambridge from a coin in the possession of Prof. Seward, F.R.S., who kindly lent it for the purpose. Lastly the view of Coniston on p. 163 is from a water-colour by Lieut. de Wesselow, R.N. kindly lent by Dr Guillemard for purpose of illustration.

1. Lancashire: Origin of the Word.

As Englishmen we are proud of our country, and we all know some of the reasons which led to the growth of the English nation and caused its people to occupy that particular tract of country which they to-day inhabit. Each of us, further, is proud of his native county. Many people of all ranks for example, young and old, take an interest in the annual struggle of the counties for supremacy in cricket. Yet comparatively few know the events which have caused our country to be separated into those divisions which we term counties. The irregular boundaries of these counties, which are so great a stumbling-block to the young student of geography, suggest that the causes which led to the making of a county are by no means simple. At the present day, when divisions of a tract of land are made, they are often very simple. Look at the line which divides Canada from the United States. For a long stretch it is straight. Many of the smaller American divisions are bounded by straight lines. So, in our country, new towns like Barrow-in-Furness and Middlesbrough are built with most of the streets in straight lines running at right angles to each other. In these cases the whole scheme of the parcelling

out is planned before the division is made. But in the
case of many of our counties there was no such principle
of arrangement. They gradually grew up under varying
conditions, and the boundaries were shifted more than
once. These boundaries have usually been determined
by some physical feature of the country which could
be readily utilised, and often formed an actual barrier

Barrow-in-Furness

between adjacent divisions. As we shall see later,
Lancashire, and also the northern part of it with which
we are specially concerned, is separated from the adjoining counties along part of its borders by hill-ridges
or by streams. Many divisions of this portion of country
which we call Lancashire were made before its present
boundaries were fixed.

All of us must have observed that the names of many counties end in "shire," as Lancashire and Yorkshire, while others, as Cumberland and Westmorland, Kent and Essex, have not this ending. Shires are tracts of land which were created by the Anglo-Saxons, the word itself being Anglo-Saxon, and meaning that it is a part *shorn* or cut off from a larger tract. The term county is from the French word *comte*, a province governed by a count (*comes*), and it did not come into use until after the Norman conquest. Such counties as Essex, Kent, and Sussex have kept their names, and roughly their boundaries as well, from the earliest times, and are survivals of former kingdoms, while Cumberland and Westmorland were not completely separated from Scotland to become entirely English until after the Norman invasion.

The name Lancashire was derived from its old capital town Lancaster, the Roman *castrum* or camp on the Lune, from whence the names Loncastre and Lancastra, afterwards changed into Lancaster. Lancastreshire was afterwards shortened into Lancashire. Professor Skeat informs me that no one knows anything about the meaning of the word Lune. The statement that it is a British word is only a guess, and there is nothing to show that it is British.

Lancashire is not a shire in the sense that it was divided from adjoining shires in Saxon times. In Domesday Book we find the lands of the southern half of the county treated with those of Cheshire, and those of the northern half with those of Yorkshire.

The boundaries of the county were practically fixed

in the reign of William Rufus, but in that of Stephen
the King of Scotland obtained possession of the territory
north of the Ribble, and it was not until the reign of
Henry II that Lancashire became definitely what it still
is, "the County of Lancaster." Lancashire then is, strictly
speaking, a county and not a shire, and it is one of the
youngest of the counties.

2. Lancashire as a whole.

In this volume we confine our attention to the
northern half of the county only, but before beginning
our task it will be well to say a few words about the
county as a whole, and the relationship of the northern
portion to the entire county.

The accompanying map shows the boundaries of the
whole county, and the relationship of that part situated
north of the Ribble, which is the subject of the present
book, to the whole. The part north of the Ribble is
shaded.

The greatest length of the entire county along a line
drawn from the Three Shire Stone at Wrynose in the
north to near Stockport in the south-east is about 80
miles, while an east and west line from Formby Point
to the county boundary east of Rochdale is over 40 miles
long.

The greatest length of North Lancashire is about 48
miles along a line from the Three Shire Stone to Preston,
while an east and west line from near Blackpool to Stony-
hurst College shows a width of about 25 miles.

Map, showing relation of that part of the county treated
in this work to the whole county

The whole county has an area of 1,203,365 acres or about 1880 square miles, that of the northern portion being approximately 462,000 acres or about 722 square miles.

The population of Lancashire by the recent census of 1911 was 4,768,474. Of these people, over 4 million inhabited South Lancashire and only about 400,000 dwelt in North Lancashire. It will be seen, therefore, that owing to the great number of industrial towns in South Lancashire and their paucity in North Lancashire the number of people per square mile is very much less in the northern division than in that south of the Ribble.

Lancashire is the sixth county in England as regards size. Yorkshire has nearly three times its area, and Lincolnshire, Norfolk, Devonshire, and Northumberland are each a little larger than our county, which is itself about one twenty-fifth of the size of all England.

The shape of the county is very irregular. That of the southern portion is more regular than the rest, being of a roughly oval outline with the longer axis of the oval extending from west-south-west to east-north-east. This part has few indentations of any size. The Ribble estuary forms a marked indent on the coast-line, and north of this river the width of the county suddenly contracts, as the eastern boundary here advances many miles westward. Owing to this and to the coastal curve of Morecambe Bay, the county south of Lancaster is very narrow, but widens again to the north of Lancaster, up the valleys of the Lune and Wenning. Near Silverdale the county of West-morland comes to the sea, and the rest of Lancashire, usually spoken of as "Lancashire north of the sands," is therefore detached from the portion which we have already considered. This detached portion is very irregular, being broken by estuaries on the south, and having a somewhat sinuous boundary-line inland.

The county may be divided into the Lancashire plain

Coniston Old Man in Winter

on the west and the high ground on the east. Much of the plain is in South Lancashire, but extends north of the Ribble where the ground west of a line connecting Preston and Lancaster belongs to it. The high ground is not continuous, but is broken up by valleys and by lowlands around Morecambe Bay. The high ground in South Lancashire along the boundary between Lancashire and

Yorkshire belongs to the Pennine Hills; to the west of these hills are minor elevations connected with them. Another tract of high ground in North Lancashire lies to the east of the North Lancashire plain. It is a nearly circular mass of which the eastern part belongs to Yorkshire. At the north end of Morecambe Bay, the Kent estuary causes the most northerly portion of the county to be absolutely detached from the southern part. This northern portion is largely highland, part of which belongs to the English Lake District.

The whole county is in contact with five other counties, namely Cumberland on the north-west, Westmorland on the north-east, Yorkshire on the east, Derbyshire on the south-east, and Cheshire on the south. The western portion from the mouth of the Duddon to that of the Mersey is coast-line.

Most of the drainage is westward into the Irish Sea, the principal river-basins being those of the Lune, Ribble, and Mersey, but about four miles south-west of Burnley a very little portion of Lancashire is drained by streams which join the Yorkshire Calder and not the Lancashire Calder, and accordingly their waters are discharged into the North Sea by the Humber estuary.

The populous nature of the southern part of the county is due to the large number of industrial centres therein, of which Manchester and Liverpool are chief. It has been already noted that but few centres of industry exist in the more sparsely inhabited northern portion.

Manchester is a cathedral city. The diocese of

Tarn Howes, Coniston

(*Lancashire in foreground, Westmorland in background*)

Manchester includes most of North Lancashire, while the rest of the northern part of the county is in the diocese of Carlisle, of which a suffragan bishop takes his title from the town of Barrow-in-Furness.

Lancashire is not only a county but a county palatine, being so made by Edward III in 1376 when he conferred the title of Duke of Lancaster upon his son, John of Gaunt, and gave him royal rights over the county where he afterwards held his court. The reigning sovereigns still retain the title of Duke of Lancaster. The duchy of Lancaster is not the same thing as the county palatine, for possessions of the duchy exist in other counties also.

The river Ribble has been selected as the dividing line between the two portions of Lancashire because it forms the southern boundary of a tract of agricultural land, which on the whole separates an industrial centre in South Lancashire, largely dependent upon the occurrence of coal, from another in North Lancashire which is in turn largely influenced by the rich deposits of iron ore.

It is true that the large town of Preston standing on the north side of the Ribble belongs rather to the South Lancashire industrial centre, but apart from this exception the division is a fairly natural one.

3. General Characteristics. Position and Natural Conditions.

North Lancashire consists of a tract of country of very varied characteristics. It lies between 53° 24′ and 54° 26′ north latitude, and 2° 28′ and 3° 3′ west longitude.

It is largely pastoral, though important industrial centres lie in the north-western portion; the large town of Preston is on its southern boundary; and the old capital of Lancaster is near its centre.

Physically the area consists; (i) of two fell regions of very different characters, the one lying to the north-west of the estuary of the Kent, and the other to the south-east of the lower part of the Lune valley; (ii) of a large expanse of low ground extending from the sea to a line drawn north and south from Hest Bank to Preston ; and (iii) of minor tracts of low ground bordering the courses of the rivers, and fringing the sea-coast of the north-western portion.

The fells differ from each other in several respects, which will be considered more fully subsequently, but three types may be noticed; the high and often rocky fells of the Furness district, the bare step-like fells of the southern part of Furness which extend eastward to Burton and Kellet, and the peat-covered moorlands of the district south-east of Lancaster.

The lowlands are very flat, mainly less than 100 feet above sea-level, with tracts of moss-land and river-flats only a few feet above high tide.

Most of the rivers are short and swift, and only

navigable in their estuarine portions, save the Lune and
Ribble, which admit small vessels some little distance
above the heads of the estuaries.

In that part of Lancashire which is included in the
Lake District is one large lake, Coniston, and a small one,
Esthwaite Water ; moreover, much of the shores of Win-
dermere form Lancashire ground, though the waters now

Preston Docks

belong wholly to Westmorland.   Several mountain tarns
also are found in the same part of the county.

The coast-line, owing to the indentation of Morecambe
Bay, is fairly extensive, and there are important ports and
watering-places situated upon it.

There are no forests, as the word is now understood,
though abundant coppice is found, especially in the

lower parts of the valleys, with thick growths of hazel, birch, willow, alder, ash, and oak, and these coppices have had an important bearing upon the industries of part of the district in past times.

The climate is mild and the rainfall rather high as compared with that of the whole of England.

The scenery of the region is varied, and much of it is very beautiful. The fell region of the higher parts of Furness is especially fine, and the moorlands to the south-east of Lancaster are impressive, and form a marked contrast to the little valleys which indent their margins.

With few exceptions the hills, owing to their generally rounded summits, are somewhat monotonous, but the valleys are in many cases very beautiful. There are miniature but picturesque waterfalls along many of the river-courses.

The scenery of the coast-line is somewhat tame, except around parts of the shores of Morecambe Bay and its estuaries, where the background of fells often helps to afford scenes of great beauty.

There is much variety in the river-valleys. The wide valleys of the lower parts of the Lune and of the Ribble especially form a marked contrast with the upland valleys of the Duddon and Leven, and there is usually much difference of detail between any two valleys.

4. Shape.   Boundaries.

When reading this chapter it will be advisable to
follow the limits of the northern part of the county with
care upon the map, and the variations in height should be
noticed, for the nature of the boundary is of considerable
importance as bearing upon the history of the area.

The length, breadth, and area of the northern part of
Lancashire have been stated in Chapter 2.

As regards its shape, it may be roughly compared to a
figure of eight lying obliquely thus ଠୄ, the northern half
of the eight, which constitutes Lancashire north of the sands,
being smaller than the southern half.   In this comparison,
however, all irregularities are disregarded: these having
been briefly noticed in Chapter 2, and an inspection of
the map will give a better idea of them than can be con-
veyed in words.   It must be remembered also that the
two halves of the eight are severed by the estuary of the
Kent.

We may now consider the boundaries, beginning with
those of the portion north of the sands.   Starting at the
Three Shire Stones the boundary follows the course of
the river Brathay to the head of Windermere.   It then
follows the west side of that lake to the foot, where it
turns up the east side, which it follows for about four
miles, then over the fells for two miles to the village of
Winster in the valley of that name.   It then descends
the Winster until it reaches the estuary of the Kent.
From the Three Shire Stones to this point the boundary
separates Lancashire from Westmorland.   The southern

Three Shire Stone

(Where Lancashire, Cumberland, and Westmorlana meet)

boundary is along the coast, and up the mid-channel of the Duddon and its tributary Cockley Beck to Three Shire Stone. It will thus be seen that except at this pass, and the little piece of fell between Windermere and Winster, the boundary is formed by water, either river, lake, or sea.

The boundary of North Lancashire lying south of the sands is much more complex. From the mouth of the Kent estuary it runs in a general easterly direction by an ill-defined and very crooked line to the Lune south of Kirkby Lonsdale. Crossing the Lune, the boundary soon reaches Easegill, up which it continues to the southern slopes of Gragreth. Along this portion also it separates Westmorland from Lancashire, but from here to the Ribble the adjacent county is Yorkshire. The boundary still remains for some distance crooked and ill-defined. It runs down the slopes of Gragreth at a very acute angle to that part lying to the north, so that a very small tongue of Lancashire lies on these slopes. Crossing the river Greeta, a tributary of the Lune, and soon afterwards another tributary, the Wenning, it ascends to the watershed of Burn Moor, part of the fell district lying southeast of Lancaster. From the slopes of Gragreth to this point its trend is nearly south. It now follows the watershed of the high moorland for many miles in a crescentic curve, the concavity of the crescent facing eastward. It leaves the watershed west of Whitewell, and descends into the valley of the Hodder, following that river southward to its junction with the Ribble.

The Ribble from this point to the sea forms the

boundary between North and South Lancashire, and from the mouth of the river to the mouth of the Kent, the boundary of the southern part of North Lancashire is the coast-line.

Skelwith Force
(*Boundary between Lancashire and Westmorland*)

5.   Geology and Soil.

Before giving further account of the physical geography of the county it is necessary to learn something of its geology, as the physical conditions are to a large extent dependent upon geological structure.

By Geology we mean the study of the rocks, and we must at the outset explain that the term *rock* is used by the geologist without any reference to the hardness or compactness of the material to which the name is applied; thus he speaks of loose sand as a rock equally with a hard substance like granite.

Rocks are of two kinds, (1) those laid down mostly under water, (2) those due to the action of heat.

The first kind may be compared to sheets of paper one over the other. These sheets are called *beds*, and such beds are usually formed of sand (often containing pebbles), mud or clay, and limestone, or mixtures of these materials. They are laid down as flat or nearly flat sheets, but may afterwards be tilted as the result of movement of the earth's crust, just as you may tilt sheets of paper, folding them into arches and troughs, by pressing them at either end. Again, we may find the tops of the folds so produced worn away as the result of the wearing action of rivers, glaciers, and sea-waves upon them, as you might cut off the tops of the folds of the paper with a pair of shears. This has happened with the ancient beds forming parts of the earth's crust, and we therefore often find them tilted, with the upper parts removed. Tilted beds are said to *dip*, the direction of dip being that in which the beds plunge *downwards*, thus the beds of an arch dip *away from* its crest, those of a trough *towards* its middle. The dip is at a low angle when the beds are nearly horizontal, and at a high angle when they approach the vertical position. The horizontal line at right angles to the direction of the dip is called the line of *strike*. Beds form strips at the surface, and the portion where they appear at the surface is called the *outcrop*. On a large scale the direction of outcrop generally corresponds with that of the strike. Beds may also be displaced along great cracks, so that one set of beds abuts against a different set at the sides of the crack, when the beds are said to be *faulted*.

The other kinds of rocks are known as igneous rocks, which have been melted under the action of heat and become solid on cooling. When in the molten state they have been poured out at the surface as the lava of volcanoes, or have been forced into other rocks and cooled in the cracks and other places of weakness. Much material is also thrown out of volcanoes as volcanic ash and dust, and is piled up on the sides of the volcano. Such ashy material may be arranged in beds, so that it partakes to some extent of the qualities of the two great rock groups.

The production of beds is of great importance to geologists, for by means of these beds we can classify the rocks according to age. If we take two sheets of paper, and lay one on the top of the other on a table, the upper one has been laid down after the other. Similarly with two beds, the upper is also the newer, and the newer will remain on the top after earth-movements, save in very exceptional cases which need not be regarded by us here, and for general purposes in our own country we may regard any bed or set of beds resting on any other as being the newer bed or set.

The movements which affect beds may occur at

different times. One set of beds may be laid down flat,
then thrown into folds by movement, the tops of the
beds worn off, and another set of beds laid down upon the
worn surface of the older beds, the edges of which will
abut against the oldest of the new set of flatly deposited
beds, which latter may in turn undergo disturbance and
removal of their upper portions.

Hawkshead

*(The town is on Silurian rocks: the distant hill is made
of Ordovician rocks)*

Again, after the formation of the beds many changes
may occur in them. They may become hardened, pebble-
beds being changed into conglomerates, sands into sandstones, muds and clays into mudstones and shales, soft
deposits of lime into limestone, and loose volcanic ashes

into exceedingly hard rocks. They may also become cracked, and the cracks are often very regular, running in two directions at right angles one to the other. Such cracks are known as *joints*, and the joints are very important in affecting the physical geography of a district. As the result of great pressure applied sideways, the rocks may be so changed that they can be split into thin slabs, which usually, though not necessarily, split along planes standing at high angles to the horizontal. Rocks affected in this way are known as *slates*.

If we could flatten out all the beds of England, and arrange them one over the other and bore a shaft through them, we should see them on the sides of the shaft, the newest appearing at the top and the oldest at the bottom. Such a shaft would have a depth of between 50,000 and 100,000 feet. The beds are divided into three great groups called Primary or Palaeozoic, Secondary or Mesozoic, and Tertiary or Cainozoic, and at the base of the Primary rocks are the oldest rocks of Britain, which form as it were the foundation stones on which the other rocks rest, and are termed Precambrian rocks. The three great groups are divided into minor divisions known as systems.

In the accompanying table a representation of the various great subdivisions or "systems" of the beds which are found in the British Islands is shown. The names of the great divisions are given on the left-hand side, in the middle the chief divisions of the rocks of each system are enumerated, and on the right hand the general characters of the rocks of each system are given.

	Names of Systems	Subdivisions	Characters of Rocks
TERTIARY	Recent Pleistocene	Metal Age Deposits Neolithic ,, Palaeolithic ,, Glacial ,,	Superficial Deposits
	Pliocene	Cromer Series Weybourne Crag Chillesford and Norwich Crags Red and Walton Crags Coralline Crag	Sands chiefly
	Miocene	Absent from Britain	
	Eocene	Fluviomarine Beds of Hampshire Bagshot Beds London Clay Oldhaven Beds, Woolwich and Reading Thanet Sands (Groups	Clays and Sands chiefly
SECONDARY	Cretaceous	Chalk Upper Greensand and Gault Lower Greensand Weald Clay Hastings Sands	Chalk at top Sandstones, Mud and Clays below
	Jurassic	Purbeck Beds Portland Beds Kimmeridge Clay Corallian Beds Oxford Clay and Kellaways Rock Cornbrash Forest Marble Great Oolite with Stonesfield Slate Inferior Oolite Lias—Upper, Middle, and Lower	Shales, Sandstones and Oolitic Limestones
	Triassic	Rhaetic Keuper Marls Keuper Sandstone Upper Bunter Sandstone Bunter Pebble Beds Lower Bunter Sandstone	Red Sandstones and Marls, Gypsum and Salt
PRIMARY	Permian	Magnesian Limestone and Sandstone Marl Slate Lower Permian Sandstone	Red Sandstones and Magnesian Limestone
	Carboniferous	Coal Measures Millstone Grit Mountain Limestone Basal Carboniferous Rocks	Sandstones, Shales and Coals at top Sandstones in middle Limestone and Shales below
	Devonian	Upper } Mid } Devonian and Old Red Sand- Lower } stone	Red Sandstones, Shales, Slates and Lime- stones
	Silurian	Ludlow Beds Wenlock Beds Llandovery Beds	Sandstones, Shales and Thin Limestones
	Ordovician	Caradoc Beds Llandeilo Beds Arenig Beds	Shales, Slates, Sandstones and Thin Limestones
	Cambrian	Tremadoc Slates Lingula Flags Menevian Beds Harlech Grits and Llanberis Slates	Slates and Sandstones
	Pre-Cambrian	No definite classification yet made	Sandstones, Slates and Volcanic Rocks

With these preliminary remarks we may now proceed to give a brief account of the geology of the northern part of our county.

In it the following systems are found and are represented on the geological map at the end of the book: Recent and Pleistocene, New Red Sandstone, Carboniferous, Silurian, and Ordovician. The figure on p. 24 shows what is called a geological section drawn across North Lancashire from the river Duddon through Morecambe Bay to Ribchester east of Preston, and gives the arrangement of the rocks. It represents what would be seen on the sides of a deep cutting if such were made along that line.

The oldest rocks form the northern part of the county belonging mainly to the Lake District. This tract extends north of an irregular line drawn through Ireleth, Dalton, Cartmel, and Lindale. The rocks are known as the Ordovician and Silurian rocks, which are amongst the oldest in the British Isles, or indeed in the world.

The Ordovician rocks consist of old lavas and ashes poured out from volcanic vents, with a band of impure limestone resting on the top of these rocks, and therefore of newer age. The tops of the Coniston Fells are composed of these rocks, and the limestone, known as the Coniston Limestone, runs along the south-east flank of those fells from the head of Windermere to Millom, and reappears, owing to a fold, on the east side of the Duddon estuary.

The Silurian rocks are composed of hardened mudstones and sandstones, the former often converted into

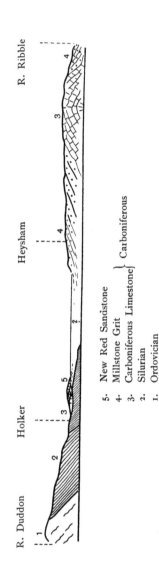

R. Duddon Holker Heysham R. Ribble

5. New Red Sandstone
4. Millstone Grit
3. Carboniferous Limestone } Carboniferous
2. Silurian
1. Ordovician

Geological Section from the Duddon to the Ribble showing the mode of occurence of the rocks of the various subdivisions, as seen along a line drawn in a general north-north-west and south-south-east direction. Length about 50 miles slates. They form the ground between the line from Windermere to Millom, and that from Ireleth to Lindale, save in the small patch on the east side of the Duddon estuary where the Ordovician rocks reappear. Owing to a great earth-fracture, a small patch of Silurian rocks appears east of Kirkby Lonsdale.

After the formation of the Ordovician and Silurian rocks they were upheaved, and their tops planed off, and accordingly the succeeding rocks of the county rest on the upturned edges of the more ancient rocks.

These newer rocks are known as the Carboniferous rocks, on account of the occurrence within them of coal. Three main divisions are found in the county, namely, the Carboniferous Limestone (or Mountain Limestone as it is often called) at the base, the Millstone Grit in the middle, and the Coal Measures at the top. Unfortunately only a very small patch of these valuable coal measures is found in our part of the county.

The Carboniferous Limestone forms an irregular strip on the north side of Morecambe Bay, and extends eastward to the county boundary on Gragreth, and owing to a minor fold, it reappears at the surface far away to the south between Garstang and Longridge. The rocks consist of a white limestone with some shales associated with them. They form prominent features on Hampsfell, Warton Crag, and other hills.

The Millstone Grit is developed on each side of the Lune from near the point where it enters the county to its mouth, but is best seen in the high moorlands which lie to the south-east of Lancaster. It consists chiefly of

Goats Water and Doe Crags near Coniston

(Crags of Ordovician rocks, with screes between crags and tarn)

massive beds of coarse sandstone or gritstone with some shales.

The Coal Measures just touch the northern part of the county near Black Burton, though the main mass, which belongs to the Ingleton coalfield, lies in Yorkshire.

After the formation of the Carboniferous rocks another movement took place, not so marked as that which followed the close of the Silurian period. The Carboniferous rocks became tilted and planed off as the Ordovician and Silurian were during the earlier movements.

Another set of rocks consisting chiefly of red sandstone was laid upon the older rocks. These are divided into two systems, an older Permian and a newer Triassic, but the rocks of both the divisions in the north-west of England are very similar, and it is convenient to speak of the rocks of the two systems under the name New Red Sandstone. They consist chiefly of sandstones and red clays. The most extensive development of these rocks is found in the tract bordering the sea between Heysham and the Ribble, to the west of a line drawn from Heysham to a point on the Ribble about four miles north-east of Preston. North of Morecambe Bay they form the Barrow peninsula and Walney Island. A small patch occurs at Cark, and another of somewhat peculiar character near the village of Westhouse towards Ingleton. The rocks themselves are usually obscured by superficial deposits.

These are the newest true rocks found in the area. Since their formation the work of rivers and glaciers has largely been concerned in cutting out the valleys, leaving the intervening portions to project as the fells. Much of

Warton Crag

(Showing "desk-structure" of a Carboniferous Limestone hill)

the work has been done by the rivers, which are able to saw their way downwards, thus deepening the valleys, while rain, frost, and the other agents of the weather cause the material of the valley-sides to be carried downwards to the streams at the base, thus widening the valleys. At a time which as compared with the formation of the rocks which we have described is but as yesterday, though remote as compared with the beginnings of the human history of our land, the district was occupied by masses of ice moving downwards from the upland regions towards the sea, and these masses of moving land-ice produced well-known characteristic marks in the shape of rocks rounded and polished by their action, which are frequent in that part of the county forming a portion of the Lake District. In addition to this, the ice helped to increase the depth and width of the valleys, and also left much of the material which it ground down and carried away in sheltered spots and lowland tracts to form the stiff clay, sometimes mixed with sand, and containing blocks of stone of various sizes, which is known as boulder-clay. This clay occupies much of the lowlands west of the line between Hest Bank and Preston. Here and there the ice left large blocks, termed by geologists "perched blocks," poised in curious positions on the sides of upland valleys. The lakes and tarns of the district occur in hollows partly due to excavation by this ice and partly to blocking of the valleys by deposits of boulder-clay or similar material, some being due entirely to one process, some to the other, and others again to a combination of the two.

Since the glacial period, the action of the weather has

Tilberthwaite Gill near Coniston: an eroded stream-gorge

caused the upper surfaces of the rocks to be broken up into pieces of various sizes, and parts of the glacial accumulations to be loosened, giving rise to soils. Of these there are the following five main types, which vary according to the nature of the underlying rocks.

In the slate tracts, the character of the soil is dependent upon the glacial accumulations which have in so many

Glacial Boulders of Silurian rocks resting on Carboniferous Limestone north-east of Cartmel

places covered the slaty rocks; where the latter are uncovered by glacial materials, they are often bare of soil. The glacial materials give rise on the whole to a poor stiff stony soil, usually wet, though where much sand occurs in the glacial masses, the soil is looser and drier.

The Mountain Limestone when not covered by glacial materials is usually bare; here and there a short sweet turf occurs. Where glacial accumulations lie thickly over the limestone, the soil naturally resembles that of the slate tracts, but where the glacial materials are thin, a fairly rich soil may be produced.

The Millstone Grit gives rise to a loose porous soil, but as much of the country formed of this rock is high fell there is comparatively little cultivable ground.

The fourth type of soil is formed over the New Red Sandstones. There is often a light sandy loam of a red colour, but on this tract also variations are produced by the presence of glacial materials.

The fifth type is found occupying the sites of former lakes which have been filled in by gravel and silt, and also portions of the estuarine tracts. On these flat areas of lands there has been as a general rule an abundant growth of peat, which yields a rich black soil. With the peat is mixed a variable amount of silt, which causes the soil to be especially valuable.

6. Surface and General Features.

North Lancashire may be divided according to its physical structure into four important divisions, two of which are upland, one lowland, and one a mixture of the two.

(1) The highest ground occurs in the Coniston Old Man group, which is part of an upland tract of High Furness composed of Ordovician and Silurian rocks. This tract also occupies parts of Cartmel.

(2) South of this is an admixture of upland and lowland forming the ground around Morecambe Bay and to near Lancaster, and extending eastward on the north of the Wenning to the county boundary on Gragreth. This consists partly of low ground formed by the alluvium of the bay and its estuaries, and of the lower parts of the river-courses, and out of it project hills of Carboniferous Limestone, the chief of which are those between Dalton and Ulverston, the fells east of Cartmel, the ground south of Silverdale, Warton Crag, and a triangular patch of country between Burton-and-Holme, Whittington, and Hest Bank. On the far side of the Lune forming part of Gragreth is a strip of high ground which belongs to the Pennines, and is the only part of that hill-group situated in North Lancashire.

(3) South of the Wenning, lying between the county boundary and a line drawn from Lancaster to Preston, is high ground forming part of a nearly circular tract of fell of which the eastern side lies in Yorkshire. Most of the Lancashire portion is composed of Millstone Grit, though the Carboniferous Limestone appears in the south, north of Longridge.

(4) Lastly we have the flat tract between the line from Lancaster to Preston and the sea, all of which, save some little patches near Kirkham, is below 100 feet, and most of it very much lower.

We will now briefly consider the general characters of these four divisions, taking them in the order stated above.

(1) The first district is bounded on the west by the

river Duddon, on the north by the Brathay, and on the east by the western shores of the upper part of Windermere, and further south by the river Winster. These all form parts of the county boundary, and beyond that boundary this fell region is continued into the Lake District of Cumberland and Westmorland. That part of it which is in Lancashire is sometimes spoken of as "Lakes Lancashire," it being actually part of the Lake District. The southern boundary of this fell region is a geological line separating the Ordovician and Silurian rocks which compose it from those of the Carboniferous Limestone which occurs further south. The line is irregular, running in a general east and west direction from Lindale on the Kent estuary, through Cartmel and Ulverston, to the Duddon estuary, north of Dalton-in-Furness.

We may subdivide this region into four minor fell-groups separated one from another by valleys and passes. These are (a) the Coniston Old Man group ; (b) the fells which extend from Torver to Ulverston between the Duddon and Crake rivers ; (c) those fells lying between Coniston and Windermere lakes on the north and between the Crake and Leven rivers on the south ; and (d) the fells separating the Leven and the south end of Windermere from the Winster.

(a) The Coniston Old Man group is the most interesting, presenting us as it does with typical lakeland scenery, and containing the most elevated tract in Lancashire. It is everywhere bounded by fairly low ground, having the Duddon valley on the west ; a tributary of the same, Wrynose Pass, and the Brathay on the north ; and Oxenfell Pass, Yewdale, the head of Coniston Lake, and a depression along which run the road and railway from Coniston village to Broughton on the south-east.

The greater part of the ground is composed of the volcanic series of Ordovician rocks, which cause the fine scenery.   Coniston Old Man, the highest fell, rises to a

Coniston Lake and the Old Man

height of 2633 feet above sea-level, but adjacent summits are little lower.   The chief are Greyfriars (2535 feet), Carrs (2575 feet), Wetherlam (2502 feet), and Doe Crags (2555 feet).   South of the latter is the high level pass of Walney Scar (just under 2000 feet) between Coniston village and Seathwaite in the Duddon valley.

A great part of this fell-region is bare and craggy.

There is little wood or coppice, but considerable stretches of peat in places.

The other three subdivisions, though distinctly fell-tracts, are much lower than the preceding, each of them rising to a height of just over 1000 feet, and much of their surface is occupied by wood and coppice. They are composed almost entirely of Silurian rocks. (*b*) That which lies between the Duddon and the Crake is divided by a north and south depression between Lowick Bridge and Ulverston, and a pass with a road runs westward from this depression to Broughton. (*c*) That between Coniston and the Crake to Windermere and the Leven consists on the west of a λ-shaped group of fells, with the Grizedale valley extending southward into Rusland Pool between the two arms of the λ. North-east of this is a depression separating the λ-shaped mass from Claife Heights on the west side of Windermere. In this depression lies Esthwaite Water. (*d*) The fourth tract is again divided by a north and south depression, the Cartmel valley, between the foot of Windermere and Cartmel. Its highest point is Gummers How on Cartmel Fell, which, though only 1054 feet high, is the highest ground which rises immediately above Windermere.

(2) Let us now turn to the area which we have defined as an admixture of upland and lowland. As previously stated, the portion around Morecambe Bay is largely composed of Carboniferous Limestone, though there is some New Red Sandstone. Much of the ground composed of the former rock rises into hills, but that made of sandstone is on the whole less elevated. These higher

areas are intersected by the estuarine flats, and contain
some mosses, which are partly of estuarine and partly of
lacustrine origin. Beginning in the west we meet with
the Furness peninsula on which Barrow is situated. This
is formed of limestone between Ireleth, Ulverston, and
Gleaston, and is there fairly hilly. South of this is a mass
of Red Sandstone forming undulating ground except near
parts of the coast, where it is much flatter, and often
covered by glacial deposits, alluvium, and blown sand, as
on Walney Island. East of this peninsula are the Ulverston sands of the Leven estuary, bordered on the east side
by mosses through which projects higher ground made of
limestone, which once formed islands. The large triangular mass of limestone between Cartmel and Grange
rises to a height of 727 feet on Hampsfell. East of this
lie the sands of the Kent estuary which, with the little
piece of land about Arnside belonging to Westmorland,
sever Lancashire into two detached portions. A mass
of limestone lies between Silverdale, Carnforth, and the
neighbourhood of Burton-and-Holme, which last place
is in Westmorland, just north of the county boundary.
The land here rises to heights of over 500 feet on
Warton Crag and in the neighbourhood of the Kellet
villages, and is still higher east of Burton. Various mosses
lie in this tract, some on the sites of former lakes, as
Burton Moss and that around Haweswater; others, as
Storr's Moss and that near Carnforth, marking former
estuaries. In the low ground between Carnforth and
Burton and also east of this up the Keer valley are remarkable ridges formed of glacial gravels. Still further

east is the ground bordering the Lune and Wenning between Burton on the north, and Lancaster and Wennington on the south. This is chiefly formed of Millstone Grit, and rises there to no great height. As before stated, however, we include here the triangular area of ground belonging to Lancashire which extends on to Gragreth. It ought strictly speaking to be put in a separate division, forming, as we said, part of the Pennine Hills, but on account of the very small part which comes into the county it is convenient to include it in the division containing the land which consists of a mixture of upland and lowland.

(3)   The third great tract is the high ground south of the Wenning, and separated from the lowlands of the Lancashire plain by the line drawn from Lancaster to Preston. It is markedly contrasted with the high ground of the Coniston group, being largely composed of rounded hills, often with comparatively flat tops, covered in many places with peat, though here and there the scarps of Millstone Grit form cliffs on the sides and summits, as is well seen on Clougha. Much of this ground is over 1000 feet high, and at Ward's Stone it rises to 1836 feet. It is diversified by considerable valleys, of which those of Roeburn and Hindburn on the north, Littledale on the north-west, Wyresdale on the west, Bleasdale on the south-west, and the western side of the Hodder valley on the south are in Lancashire. Belonging to this area is the somewhat isolated mass of Longridge Fell, near Longridge, which rises to a height of 1149 feet, and is separated from the main mass of fells by the comparatively low ground in which Chipping lies.

(4) The last great tract, lying between the line from Lancashire to Preston and the sea, is the least diversified. It forms the northern part of the great Lancashire plain. Save for a little Millstone Grit about Heysham it is entirely formed of New Red Sandstone rocks, which however rarely show at the surface, for the country is usually thickly covered with glacial accumulations. These often form rounded ridges, which to some extent break the monotony of the scenery. Other parts are mosses, as Pilling Moss; along the sides of the estuaries are tracts of salt-marsh, while tracts of blown sand occur about Fleetwood and Lytham. The comparatively high ground between Broughton, Kirkham, and Weeton is the watershed between the Wyre and the Ribble. The watersheds between the Lune, the Conder, the Cocker, and the Wyre on this low tract are badly defined, and these rivers wander hither and thither seeking an escape to the sea.

7. Along the Coast.

The coast-line may be regarded as beginning near Foxfield Junction, Broughton-in-Furness, where the Furness railway crosses the estuary of the Duddon. From here to the mouth of the Ribble the distance measured in a straight line is 38 miles. But the actual length of coast, measured around Morecambe Bay, and omitting the portions up the estuaries of the Leven, Kent, and Lune, and the minor indentations, is over 70 miles.

We will follow the coast-line from near Broughton-in-Furness to the mouth of the Ribble, noting the principal features on the way.

From Foxfield Junction for about five miles the coast runs on the east side of the Duddon estuary nearly south to Askham, along low ground backed by hills of slaty rocks from a mile to a mile and a half inland.   The low

Barrow-in-Furness: Walney Ferry

ground is largely moss, the principal tract being Angerton Moss.   Some way north of Askham the former island of Dunnerholme, a rocky mass of mountain limestone, rises out of the alluvium, and projects into the estuary.   A little south of Askham a promontory of blown sand, Sandscale Hawes, projects into the estuary, and at its southern end is the channel which, turning south, separates the island called Walney from the mainland. This island is about 10 miles long, and is largely covered with blown sand and alluvium, once almost entirely warren, but now becoming built over by the extension across the narrow Walney channel of Barrow-in-Furness. Four smaller islands occur on the east side of the south end of

Humphrey Head

Walney, namely Sheep Island and Piel Island on the west side of the channel and Foulney and Roe Islands on its east side.

Barrow itself with its docks has a frontage of upwards of two miles on the east side of the channel.

Hilpsford Point at the south end of Walney forms the northern entrance to Morecambe Bay, and the description

of the coast from here to Rossall Point is an account of
the borders of that bay.

From Rampside at the entrance of the Walney
channel the coast turns to the north-west and forms
low ground to Newbiggin. The whole coast between
Sandscale Hawes and this village is composed of Triassic
rocks and is bordered by fairly low-lying ground. From
here to Cark on the other side of the estuary the
Carboniferous Limestone forms the coast except where
overlain by the alluvial flats of the Leven estuary
between Conishead Priory and Cark. Several rocky
eminences, formerly islands, rise out of this alluvium
on either side of the channel of the Leven, which
is crossed between Ulverston and Cark by a viaduct of
the Furness railway. South of Cark, a mass of Triassic
ground flanked on the east by a marsh extends into the
bay, and east of the marsh is the elevated promontory
of Humphrey Head, formed of Carboniferous Limestone,
which borders the coast from this point to Warton Crag,
except where the waters of the Kent flow through between
Grange and Arnside. The coast is fairly high at Grange,
a sheltered watering-place facing east, with the rocky
Holme Island lying off it, situated close to the county
boundary, which here comes south down the Winster river.
The viaduct of the Furness line over the Kent estuary
is in Westmorland, for the county boundary strikes across
the sands, and accordingly the first strip of coast met with
on the east side of the estuary is in Westmorland. We
soon re-enter Lancashire, traverse about two miles of
rocky coast with abrupt cliffs, cross an alluvial tract, and

find ourselves under Warton Crag, where a low promontory, Ings Point, is made of Millstone Grit, let down
against the Carboniferous Limestone of the Crag by a
gigantic fault.

At Carnforth, with the high chimneys of its ironworks, the river Keer enters the bay, and the coast, which
from Grange has had a south-easterly trend, here turns

Grange-over-Sands

south-west. The ground now becomes low, entering
the great lowlands of West Lancashire, though two
imposing cliffs of red boulder-clay are met with between
Carnforth and Bolton-le-Sands. Hest Bank is a little
watering-place, beyond which the coast enters an alluvial
tract, on which stands the large watering-place of Morecambe, three miles beyond which is Heysham, which has

recently become important as the port of the steamers of the Midland Railway Company which run from here to Ireland.

South of Heysham the New Red Sandstone rocks set in and occupy the rest of the coast-line, though generally covered by glacial deposits or alluvium. The coast once more runs south-east, for a semi-circular minor bay extends from here to Fluke Hall near Pilling. Into its northern sweep enters the Lune through its estuary, with Sunderland Point on the north-west and Cockersand Abbey on the opposite shore, while Glasson Dock is about two miles and Lancaster about seven miles from the mouth of the estuary. The river Cocker enters this bay at Cockerham. At Fluke Hall sandhills are met with and extend to Rossall Point, being cut through by the estuary of the Wyre, with Knott End and the port and watering-place of Fleetwood at its mouth, the former on the right and the latter on the left bank. Rossall Point forms the southern termination of Morecambe Bay.

South of Fleetwood the coast is nearly straight for twelve miles. It is formed of alluvial deposits for four miles and then by somewhat higher ground with cliffs of boulder clay for a similar distance to the watering-place of Blackpool, which extends for about three miles along the sea-front. The watering-place of St Anne's-on-Sea is situated at the mouth of the estuary, and that of Lytham about three miles to the east. From the former place the coast begins to curve into the estuary of the Ribble, and is bordered by blown sands to Lytham. From Lytham the estuary extends eastwards towards Preston, which is about ten miles from Lytham.

Ironworks : Carnforth

A few words must be added about Morecambe Bay. At high tide the water fills the bay, and runs up the estuaries, the phenomenon known as the tidal bore being a feature in the estuarine tracts. At low water, however, not only are the floors of the estuaries exposed, the rivers being confined to narrow, often shifting channels, but large areas of the bay itself are bared, showing the Lancaster Sands. These, as the name implies, are largely sand, but much mud is also brought down by the rivers, and here and there along the shores are patches of large blocks of stone called skears.

Four principal channels occur, those of the Leven, Kent, Keer, and Lune, the first three being directed towards the south, and the last towards the west, but other depressions are found marking the sites of abandoned channels. From Hest Bank to Kent's Bank is the old passage of the sands which crossed the channels of the Keer and Kent. The traverse of this passage was accompanied by frequent loss of life.

The entrance to Morecambe Bay is marked by two lighthouses, one at Hawes Point at the south of Walney, the other off Fleetwood, and a lightship is moored in the centre of the bay, four miles west-south-west of Heysham. A lighthouse north of Cockersand Abbey marks the entrance to the Lune estuary. Another lighthouse at St Anne's-on-Sea marks the north of the entrance to the Ribble. The lighthouses all round our coasts are built and supported by a branch of the Civil Service known as Trinity House. The Elder Brethren of Trinity House obtain the funds for the purpose by levying light dues on the ships which enter and leave British ports. Minor lighthouses mark the harbours.

The coast-line has undergone various changes in historic times. Parts of the coast are worn away by erosion, while in other places addition is made by de-

Blackpool: High Tide

position of shingle and sand. A great part of the coast of Morecambe Bay (especially of its north-western extent) has suffered loss by erosion, and some villages have been thus destroyed. South of Morecambe Bay, erosion has occurred from Rossall to a point south of Blackpool: from thence towards Lytham actual addition of land has resulted from deposits.

It may here be noted that some of the old salt-marshes along parts of the coast, especially in the estuarine tracts, have been recently reclaimed by artificial means, and converted into rich agricultural ground.

8. Watersheds and Passes.

The main watersheds separate the principal rivers of the area; these we may notice in order, proceeding from north-west to south-east.

The north end of the Coniston group of fells forms the watershed between the Duddon and the Brathay, while its southern portion separates the drainage of the former river from that of the Crake. The ridge of the line of fells between Torver and Dalton also separates the Duddon drainage from that of the Crake, but in its northern part only, while at the south end the drainage from this ridge goes into the estuaries of the Duddon and Leven respectively. The fells about Cartmel form a watershed between the Leven drainage and that of the Winster. The high ground about Kellet sends its drainage into the Keer on the north and to the Lune on the south-east. The northern side of the great mass of upland east of the line from Lancaster to Preston drains into the Lune, while its west and south-west sides drain into the Wyre. Lastly, the isolated Longridge Fell forms a watershed between the Wyre and the Ribble, and this shedding line is continued on comparatively low ground to the west past Kirkham.

The passes actually situated within the county are of minor importance as affecting the original immigration of the various peoples who came into the district, for they arrived chiefly along the low ground from the south, or else entered from the sea. There is no high ground in the district which absolutely divides two tracts of low ground desirable for settlements.

In the lakeland portion of the district some of the passes are important to tourists, as, for instance, the Oxenfell and High Cross Passes, on the roads between the heads of Windermere and Coniston Lakes. That of Wrynose, between the Brathay and the Duddon valley, now mainly a tourist route, was, as we shall ultimately see, of importance at one time as permitting the passage of a Roman road. On its summit stands the Three Shire Stone marking the boundary between Lancashire, Westmorland, and Cumberland. Other passes serve for the purpose of allowing pack-roads to traverse the ridges at levels lower than those of the hill tops. Along these produce was carried from valley to valley by horses, when the valley-bottoms were barely passable owing to dense growth of coppice. An example is the Walney Scar Pass between Coniston village and the Duddon valley. These tracks are abundant in the lakeland part of Lancashire, but their use has been largely superseded by that of better roads along the lower levels. Many of these roads are carried over minor passes, which the reader may study on the map at the beginning of the book.

On the high land east of the line between Lancaster and Preston, three passes connect the lowlands on the

north and west with those on the south and east, all
being situate on the county boundary between Lancashire
and Yorkshire. The principal, which is also the most
southerly, is the Trough of Bowland, through which
passes the road between Lancaster and Clitheroe.
Another, about four miles north-east of this, allows of
communication between Hornby and the Hodder valley ;
and a third, three miles to the north-east, has a road
which goes between Bentham (in Yorkshire) and the
same valley. Between the last two passes the old Roman
road went from Ribchester to Overborough, near Kirkby
Lonsdale, but it seems to have been carried boldly over the
ridge-summit.

Lastly, some of the river-valleys lead to passes which,
although outside the county, are of importance as permitting communication between dwellers in the county
and those outside. The low grounds along the Lune
above Lancaster, and those between Warton Crag and
Burton, lead up towards the Shap Pass in Westmorland,
giving communication with the lowlands of northern
Westmorland and Cumberland, and so with Scotland.
The Lune valley above Lancaster and the Wenning
valley allow of communication with the Ribble valley by
a low pass near Giggleswick, and as the Ribble valley is
near Hellifield separated from the Aire by another low
pass, there is ready access between the country around
Lancaster and Preston and the populous part of the West
Riding of Yorkshire. These routes have been of great
importance from very early times, as will be seen when
the Roman roads are described.

9. Rivers.

The rivers of North Lancashire flow into Morecambe Bay, with the exception of the boundary river—the Duddon—on the north-west, and the Ribble, which we have taken as our division between North and South Lancashire on the south.

The Leven, below Windermere

We may consider them in order, beginning with the Duddon.

The river Duddon rises at Three Shire Stone where Cumberland, Westmorland, and Lancashire join, and it forms the boundary between Lancashire and Cumberland from source to sea, a distance of about 15 miles. It is

essentially a Lakeland river, flowing through the Ordovician volcanic rocks from its source to the estuary. The
waters are swift and clear, and the hills rise to considerable
elevation on either side of the river-banks. The estuarine
portion has been noticed in the chapter treating of the
sea-coast.

The Crake, flowing from Coniston Lake, joins the
Leven at Greenodd, about six miles below the point
where it quits the lake.

Proceeding eastward, the next river of importance is
the Leven, which flows from the foot of Windermere to
its estuary. Above Windermere two important feeders
belong to the drainage of the Leven basin. One, the
Rothay, is in Westmorland; the other, the Brathay, rises
at Three Shire Stone in the direction opposite to that
taken by the infant Duddon, and forms the county
boundary between Lancashire and Westmorland from its
source to its termination in Windermere.

The Leven itself below Windermere flows through
a narrow valley for about four miles to the village of
Haverthwaite. There it enters the alluvial flat at the
head of the partially silted-up estuary, and about Greenodd, a little lower down, the estuary proper begins. The
scenery of the estuarine part of the Leven is very fine.

Proceeding again eastward we reach the estuary of
the Kent. This river is in Westmorland, but a portion
of the estuary near Grange-over-Sands is in Lancashire,
and the Winster, which enters the estuary just above
Grange, forms the county boundary between Westmorland and Lancashire along the greater part of its course.

The Keer rises south of Hutton Roof and for a short distance forms the county boundary. After a course of a few miles it enters Morecambe Bay at Carnforth. Though the river is small, it flows through a large valley, and there is reason to believe that in times before the glacial period the Lune itself found its way to the sea along the lower part of this valley.

The Lune: view from Aqueduct, Lancaster

The Lune rises far away north of the county boundary on the northern slopes of Howgill Fells in Westmorland. About 30 miles from its source it enters Lancashire territory just south of Kirkby Lonsdale. From that point to the sea at the mouth of the estuary is a distance of a little over 20 miles. The views in the Lune valley between Kirkby Lonsdale and Lancaster are very beautiful; in many places the great mass of Ingleborough is seen in the background, when looking up the valley. Near Tunstall the Greeta, flowing from the east, joins the Lune, and below Hornby the more important Wenning, which rises near Clapham in Yorkshire. The Wenning itself receives the united waters of two streams, the Hindburn and the Roeburn, which rise on the Millstone

Sandel Holme on the Hodder

Grit fells east of Lancaster, and flow northward to unite at Wray. From Tunstall to Caton the river flows through a wide valley with alluvial floor, but below Caton the valley narrows to a point just above Lancaster, and on leaving the alluvial flat flows in a loop, the celebrated "Crook of Lune." At Lancaster the river becomes tidal, and no great distance below that town the estuary begins.

The Conder rises near Caton. Like the Keer its source is close to the Lune, and the watershed between is low. The stream flows a little west of south to Ellel, and then turns north-west to enter the estuary of the Lune at Glasson Dock.

The Wyre has its source towards the centre of the high mass of Millstone Grit fells near the Trough of

Lower Hodder Bridges

Bowland. It flows westward in an upland valley to Dolphinholme, where it enters low ground and turns first south to Garstang and Catteral, then south-west to St Michaels, west to near Poulton, and lastly north-west to Fleetwood, where it enters Morecambe Bay. Leaving out of account minor windings the total length of its course is about 25 miles.

The Ribble rises in Yorkshire at Ribblehead, ten miles north of Settle. About 24 miles from its source it reaches Lancashire and forms the boundary between Lancashire and Yorkshire to the river Hodder, where it flows entirely in Lancashire, and at this point begins to form the boundary of that portion of Lancashire with which we are concerned. The Hodder rises in the circular mass of Millstone Grit fells in Yorkshire, and touches Lancashire above Whitewell, forming the county boundary from there to its junction with the Ribble.

The Ribble flows through fairly high country from its source to the neighbourhood of Preston, though the valley itself is in most places wide. Above Preston it enters low ground, soon becomes tidal, and at no great distance below Preston the estuary proper begins. The estuarine portion is about 12 miles in length, and the whole length of the river from source to sea, omitting windings, is over 50 miles.

10. Lakes.

It was remarked in Chapter 3 that the only large lakes situated in Lancashire are Coniston and Esthwaite, but as a great part of the shores of Windermere are Lancashire soil, we must say something about this lake.

Windermere has a length of $10\frac{1}{2}$ miles, and covers an area of $5\frac{7}{10}$ square miles. Its maximum breadth is just under a mile opposite Millerground Bay, and the average breadth just over half a mile. It is 130 feet above sea-level,

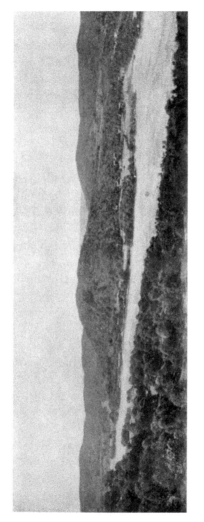

The Foot of Windermere

and drains an area of nearly 90 square miles, most of
which is in Westmorland. The greatest depth is 219
feet, at a distance of a mile and a half from the lake-
head. The lake runs nearly north and south, but the
upper part bends slightly westward towards the head.
The shores are not greatly indented, Pullwyke, near the
head on the western side, being the most marked bay.
Many islets or "holms" occur, but all are in Westmorland.

Esthwaite Water is a smaller lake between Windermere and Coniston, near the little town of Hawkshead. It
is about 1½ miles in length, and less than half a mile wide,
with a depth of about 80 feet. It lies nearly north and
south, and from it flows a small beck—Cunsey Beck—
into Windermere.

Coniston Lake is almost parallel with the lower
part of Windermere, and lies nearly north and south.
Its length is nearly 5½ miles, and it has an area of a little
under two square miles. Its greatest width is just under
half a mile. It is 143 feet above sea-level, and drains an
area of 23 square miles. The greatest depth is 184 feet near
the centre of the lake. The sides approach straightness,
and there is only one important bay, north of Coniston
Hall. Two small islets occur on the east side, close to
the shore: Fir Island, about half-way down, is a low flat
pile of stones nearly touching the shore, while Peel Island,
towards the foot, is formed of well-glaciated rock rising
out of deep water. The head of the lake is being filled
up by the alluvium brought down by a small stream, and
about a mile south of the head a large delta has been built

by Church Beck on the west side of the lake, and still
lower on the same side is the delta of Torver Beck.

North Lancashire possesses several tarns. The term
tarn is generally applied to a small lake, usually less
than half a mile in length. Many tarns lie in hollow
combes on the hillsides of the Old Man group far above
the floors of the main valleys, and in many cases the

Windermere: the Ferry, from the Ferry Nab

streams which come from them flow in cascades down
the sides of the larger valleys. Other tarns are situated
in the lowlands. In upland and lowland alike there were
formerly many more which have now become filled up,
and their sites converted into peat-bogs. The tarns are all
situated in the northern part of the district, except the
little Marton Mere, east of Blackpool. Five occur in
combes on Coniston Old Man and the adjacent fells,

namely, Levers Water, Low Water, Goats Water, and
Blind Tarn, on the western side, and Seathwaite Tarn on
the east. The last-named, which was the largest, has
now been converted into a reservoir for Barrow-in-Furness.
Low Water, notwithstanding its name, is at a great
altitude, being 1786 feet above sea-level, while the little

Urswick Tarn and Village

Blind Tarn, below Doe Crags, which has no stream
issuing from it, is 1842 feet above the sea. A small pool,
Boo Tarn, on the road from Coniston to Walney Scar,
has an exit at each end.

Two little upland tarns north of Coniston Waterhead
have been raised to form one artificial lake, Tarn Howes,
celebrated for the beautiful views obtained therefrom.

Two somewhat similarly placed lakelets, High and Boretree Tarns, lie west of the south end of Windermere.

Of valley tarns in the Lake District portion of the county are Little Langdale Tarn and Elterwater in the course of the river Brathay, along which the county boundary runs; Blelham Tarn south of Pullwyke Bay, Windermere ; and two pools, one near the head and the other near the foot of Esthwaite, of which they once formed a portion, being now separated from it by alluvium.

In the lowlands mention may be made of three tarns, Urswick Tarn near the village of that name, Haweswater near Silverdale, and Marton Mere near Blackpool. In addition there are some smaller sheets, scarcely more than pools.

11. Scenery.

In a district of which the scenery is an important factor affecting the inhabitants, some attention must be paid to its causes and character. These are dependent partly upon the geological structure, partly upon meteorological conditions, whether acting directly—e.g. the effects of sunlight and clouds upon the view—or indirectly—as affecting the vegetation, and partly also upon agents such as frost, rivers, and glaciers, by which the details of the scenery have been largely determined.

At the outset we may take into account the effects of

the more important rock-groups in controlling the nature of the surface.

The volcanic rocks of the Ordovician system are responsible for the wildest scenery in the county. They are the hardest of the rocks which are extensively found, and as there is considerable variety in their hardness, the lavas and many of the volcanic ashes being peculiarly hard, while some of the ashes are softer, a considerable diversity of outline is thus caused by these alternations. Again, they are affected by very regular systems of gigantic cracks or joints, often with belts of smashed rock along the cracks, and these cracks have been lines of weakness which have frequently been worn into notches and gorges, and they also define the sides of cliffs. To the hardness of the rocks we owe in a great degree the superior elevation of the fells which are composed of them ; and to the variations of hardness and the nature of the joints are due the frequent alternations of cliff and slope which are so marked a feature of these fells. These alternations are well displayed in the hills of the Coniston group.

The Coniston Limestone is too narrow to produce important scenic effects.

The Silurian rocks give rise to tamer scenery, for the rocks are not so hard as those of the volcanic group, and there is less variety in the degree of hardness. Accordingly they rise on the whole to less elevation, and though crags are numerous, they are on a smaller scale than those of the rocks just noticed, while the peat-covered uplands are more frequent. This type of scenery is well displayed in the country between Windermere and Coniston.

The Carboniferous rocks produce two very different types of scenery, one being characteristic of the Carboniferous Limestone, and the other of the Millstone Grit.

The structure which is presented by the mountain limestone hills has well been named "writing-desk structure," for the gently-inclined beds form gentle slopes, with steep cliffs determined by the nearly vertical joints on the other sides of the hills. This structure is typically shown in Warton Crag and Hampsfell (p. 28). The bare white cliffs and fissured "clints" offer a marked contrast to the surfaces of the fells formed out of the slate-rocks.

The clints just mentioned are flat or fairly flat surfaces of limestone with fissures or "grikes" produced by the widening of the vertical joints by acidulated rain-water. The water is capable of dissolving the limestone, and the bare flat or gently-sloping limestone surfaces are therefore often traversed by two sets of fissures at right angles to each other, penetrating for many feet or even yards from the surface. The sides of these are often honeycombed by the solvent action of the rain assisted by the vegetation that may grow abundantly within. Such clints are seen on the higher parts of the two fells mentioned above.

The massive well-jointed Millstone Grit, when found on fell-tops and to some extent on fell-sides, produces steep scarps like those seen on the top of Clougha and other parts of the fells lying south-east of Lancaster, but the abundant growth of peat often covering glacial accumulations of these fells has prevented the extensive development of such scarps, and the Millstone Grit tract of these high uplands is chiefly marked by somewhat

Wetherlam in Winter

dreary moorland, though it is beautiful during the period when the heather is in bloom.

The rocks of Triassic age, being soft, are readily worn away, hence the ground occupied by them in the country west of the line between Lancaster and Preston is low. Furthermore, their rapid breaking up allows abundant formation of soil, and the bare rock is rarely exposed. In addition to this, a great deal of glacial material has accumulated on this low ground, and masked the rocks beneath.

The effect of the ice of past ages in hollowing the valleys has been noticed in the geological chapter. We are here concerned with its deposits. We may first notice the little moraines which were left by the upland glaciers among the hills of the Coniston group. They consist of hummocky hillocks of clay, gravel, and stones, covered with coarse vegetation. A very perfect example blocks the Blind Tarn, beneath Doe Crags, Coniston.

The boulder-clay, which was noticed in the geological chapter, is spread widely over the low-lying grounds. It is often arranged so that its upper surface forms parallel mounds like the backs of whales. These are known as "drumlins," and are well seen near Lancaster. Somewhat similar mounds, but composed of sand and gravel, are termed "eskers." They occur in the neighbourhood of Carnforth.

Latest of the deposits which have produced large effects upon the scenery are those which fill in lakes and estuarine tracts. We have seen that Coniston is being filled in by delta growth where rivers enter the lake.

Many old lakelets scattered over the district have been thus filled. Burton Moss (lying partly in Westmorland) is an example.

The estuaries of the Duddon, Leven, Kent, and Lune have been silted up for long distances from their heads. Three tracts of deposit can be made out. At the old heads are the most ancient deposits, often largely

Esthwaite Lake

covered with rich peat, which gives rise to a rich fertile soil. Seaward from this we may meet with reclaimed marshes, which are still occupied by a growth of salt-marsh plants, and lastly we find the sand-banks at the mouths of the estuaries still covered by tide at high water.

Looking over these flats on the north side of More-cambe Bay we get a combination of three types of scenery, that of the Carboniferous Limestone hills with the marshes below, and the hills of Silurian and Ordovician rocks beyond. Such views are especially fine.

In the district south of Morecambe Bay are Pilling, Cockerham, and Winmarleigh Mosses, having a somewhat similar origin.

We may now refer briefly to some of the minor features.

The existence of sand dunes along parts of the coast-line of the Duddon, and again near Fleetwood and St Anne's-on-Sea, has already been mentioned. They are formed by on-shore winds driving up the sands of the shore. They are often largely covered with spear-leaved vegetation, and produce a somewhat remarkable effect.

The influence of the lakes and tarns upon the scenery need merely be noticed in passing. We have said much of these sheets of water in earlier sections.

The accumulations of loose blocks detached by the weather from the cliffs above to settle in fan-like forms on the slopes beneath have a considerable effect upon the character of the hill-sides. These "screes," as they are termed, are more extensively developed in Cumberland, but they are quite abundant in North Lancashire also. They are well seen below Doe Crags near Coniston Old Man, where they have blocked the end of Goats Water.

Many gorges have been hollowed out by stream action since the occupation of the county by ice, especially along the principal joints and the belts of smashed rock among the hills of the volcanic group of rocks. The finest of these is Tilberthwaite Gill near Coniston. They often, as at that place, contain waterfalls. The waterfalls of

North Lancashire are remarkable for their beauty rather
than for their size.   Some of them occur in the gorges as
just noted, others are rather cascades where the waters
from the upland combes pour down into the main valley,
as those from the combes containing Levers Water, Low
Water, and Seathwaite Tarn.   Others again occur where
hard rocks are found overlying soft rocks along the courses
of streams.   Such are found in some of the valleys flowing
into the Lune from the Millstone Grit Fells.

The precipices of the district are small.   The principal
are found among the volcanic rocks of Ordovician age,
the most marked being Doe Crags.   They are also found
on the Carboniferous Limestone hills, and to some extent
in the scarps of the Millstone Grit country.

In the limestone tracts some remarkable features are
caused by the tendency of water containing carbonic acid
to dissolve the lime.   The formation of grikes has already
been noticed.   When streamlets run from clayey ground
on to the bare limestone, they rush down the fissures, and
working along the bedding planes of the limestone may
excavate caverns along the latter.   A well-known cavern
of this type is Dunald Mill Hole near Nether Kellet; the
stream which here enters the cavern has an underground
course of about $2\frac{1}{2}$ miles, and emerges near Carnforth.
Two caves, Dogholes on Warton Crag and Kirkhead
Cave near Cartmel, are of interest on account of the
relics of animals and man which they have yielded.

The influence of vegetation on the scenery will be
noticed in the chapter treating of Natural History, and as
for the atmospheric effects, one need only remark that the

variability of the climate, which is sometimes treated as a matter of regret, is responsible for scenic effects which are far more beautiful than would be the case were the climatic conditions of a more settled character.

12. Natural History.

Botany and zoology are the sciences which treat of the world's flora and fauna, but the study of the distribution of plants and animals—where they are found and why—forms part of the domain of geography, for from it we learn many facts concerning the past history of the land. Every one knows by sight a certain number of the plants and animals of his own county, and this knowledge will enable him to get some idea of the way in which their geographical distribution is effected.

Let us in the first place consider the plants of the county. Some of these are commonest in the south of England, others in the east, and others again in Scotland, while a very large number of the whole are spread over the entire island, and a few are very local, so far as our country is concerned.

These plants have not originated where they now grow. We have seen in the geological section that the district was once occupied largely by ice. At that time a few plants may have lived on the rocks of the higher fells, just as they do on the hills appearing above the ice of Greenland at the present day. But, as the ice receded from the county, places must have been left bare on which plants gradually sprang up, as their seeds were wafted by the wind, or brought by birds, or in some other manner, from other regions.

It must not be supposed that the plants which were thus brought came from the regions where they are now commonest in Britain. There are many reasons for believing that at no remote geological date, though before the beginning of historic times, England was joined to the Continent to the east and south. This would afford a ready route along which the plants which undoubtedly reached England from the Continent could gradually migrate, just as, at a later period, successive immigrations of people came along that route, having only to cross the narrow straits. And as the more barbarous people were driven into the mountain fastnesses by their more highly civilised successors, so might the early plants be replaced by others which, under altered climatic conditions, were able to flourish.

But not all the human immigrants into Britain came by way of the narrow passage of the Straits of Dover. The sea-faring Danes and Norsemen, for instance, landed sometimes on the north-east and even on the west coast of England. Similarly some of our plants may have come in along some other route when England was united to the Continent not merely by the land now occupied by the straits, but by land masses which once existed over part of the site of the North Sea.

Certain plants now well established have been introduced by man. Most noticeable among these are such as grow in cornfields, which have been accidentally brought into the country together with the corn. Some have been recently introduced and are not yet established, as the small toad-flax, others like the large blue speedwell (*Veronica Buxbaumii*), though of recent introduction, are now thoroughly established, and some, like the blue cornflower, have been so long inhabitants of the country that the period of their introduction is unknown.

It will be seen from these remarks that the question as to the mode in which our county became stocked with its plants is very complicated, and as it requires much knowledge of science to sift the evidence, this part of the study of distribution is only for those who are possessed of considerable botanical knowledge.

There are other facts connected with distribution however which can readily be tested. It will soon be found that the plants of Lancashire do not flourish equally in all parts of the county, for instance, those growing on the flats near the sea at the mouth of the Lune are very different from those which live on the higher parts of Coniston Old Man. Very little observation will show that there are two important causes of this difference among the plants of various tracts of the county, namely height above sea-level and difference of soil. Let us first regard difference of altitude. Many plants are confined to tracts less than 900 feet above sea-level, of which the gorse or whin is a good example. Above 1800 feet the bracken practically ceases. Lastly, in the belt between 1800 and 2700 feet we find a remarkable assemblage of plants of an alpine character, such as the alpine campion (*Lychnis alpina*) and the rose-root

(*Sedum rhodiola*).   The plants of this belt are found in
the Coniston hills, but are much more abundant in those
parts of Lakeland which belong to the adjoining counties.
It will be a useful exercise for the student to discover for
himself what are the upper limits of the various plants
with which he is acquainted, though he must be prepared

Geranium sanguineum, **variety** *lancastriense*

to find an occasional straggler above the height to which
the species as a whole ascends.

An easier study is that of the distribution of plants
according to the soil, it being remembered that this soil
in many cases varies in character owing to the nature
of the rock beneath, though some soils, as those formed of
peat, are largely independent of the underlying rock.

Some plants are confined to the muddy silt of the salt-marshes by the sea-shore. A conspicuous example is the purple sea-aster or starwort with yellow eye (*Aster tripolium*), which grows on the salt-marshes of the estuaries around Morecambe Bay. Other sea-plants are found on sandy or gravelly soil. Special mention must be made of a geranium (*Geranium lancastriense*), found on Walney Island and nowhere else.

The bog-plants live in bogs at various heights from sea-level to the tops of some of the highest hills. Among these are the louseworts (*Pedicularis*), the insect-eating plants known as butterwort (*Pinguicula*) and sundew (*Drosera*), and handsomest perhaps of all, the grass of Parnassus (*Parnassia palustris*). In the pools among the bogs we find other plants as the bladderworts (*Utricularia*) and the pale blue-flowered water-lobelia (*Lobelia dortmanna*).

Some plants are confined to the rich soils along beck sides, as the globe-flower (*Trollius europaeus*) and the yellow balsam (*Impatiens noli-me-tangere*), the latter a truly local plant.

In the rough pastures we may find many kinds of orchis, with other plants too numerous to mention.

In the limestone district is a group of plants which flourish, notwithstanding the general dryness of the tract. Among such are the centaury (*Erythraea centaureum*), the rock-rose (*Helianthemum vulgare*), and the lady's finger (*Anthyllis vulneraria*). But the most noticeable plants of the limestone tract are those which live in the fissures of the cliffs and clints, such as the hart's-tongue fern and the yew.

Before leaving the consideration of the distribution of plants there is one matter concerning which a few words must be said.

We saw that at over 1800 feet a number of plants were found which, with us, do not occur below that height. These however are widely scattered in European mountain regions, many being found on the Scotch hills, the Alps of Switzerland, the mountains of Norway, and on lower ground within the arctic circle. They are at the present day mainly characteristic of alpine and arctic regions, and it is believed that they became established in our country during the glacial period, occupying then the British lowlands, just as they now live on the lowlands within the arctic circle. As the climate grew warmer they were displaced by other plants which were able to flourish to so great an extent as to exclude these "alpines," which accordingly were driven higher and higher, and are now found obtaining here and there a precarious footing upon our higher fells, from which perhaps they are doomed to disappear at no distant date. Let us hope that the disappearance, if it comes, will be natural, and not quickened by the wanton removal of the roots of the plants by the too eager collector.

In a county of which the scenery has within recent times had a marked effect upon the dwellers therein, a few words may be added as to the effect of the plants upon that scenery.

Many plants grow in sufficient number to produce a striking influence upon the view. The flowers of the ragwort in the rough pastures and the curious growth of the cotton-grass when in seed, may be cited as examples. There are, however, two plants whose influence is particularly pronounced, namely the heather and the bracken. The effects of heather are most striking on the moors to the east of Lancaster. The amount of heather in the Lakeland portion of Lancashire is comparatively small, but it is in this tract that the effects of the bracken are so fine. Of it Wordsworth speaks thus:—"About the first week in October, the rich green, which prevailed through the whole summer, is usually passed away. The brilliant and various colours of the fern are then in harmony with the autumnal woods : bright yellow or lemon colour, at the base of the mountains, melting gradually, through orange, to a dark russet brown towards the summits, where the plant, being more exposed to the weather, is in a more advanced state of decay."

About the animals of the district we need say less. Gifted with power of locomotion, their distribution is as a whole wider than that of the plants.

The larger quadrupeds have disappeared from the county. In prehistoric times, the north of England was occupied by lions, hyaenas, and elephants, but they were extinct in this county before man had arrived in Lancashire, though the south of England was occupied by man at the time that these beasts lived there.

In historic times the wild boar, wolf, red deer, and wild white cattle existed in Lancashire. It must be understood that the Forest of Bowland was then true forest, forming a resort suited to these large mammals. The wolf disappeared from England about the time of

Henry VII, and one of its last retreats was the above-named forest. Wild boars are recorded at Hoghton Towers near Blackburn in 1617, and they no doubt extended to the north of the Ribble. The wild white cattle of the same place became extinct about two hundred years ago, and the last herd of red deer in Bowland was destroyed in 1805.

The fox is yet found on the hills, and the otter in the streams, and fox-hunts and otter-hunts are still exhilarating pastimes. The badger is getting very rare, and the wild cat, abundant a hundred years ago, is almost certainly extinct.

Of marine mammals, whales and seals are occasionally recorded, and porpoises are more common.

Of the birds, the eagle has disappeared, but a number of different kinds of hawk are found, especially in the fell-districts.

The buzzard has nearly gone and the merlin is rare. Sparrow hawks are fairly common and kestrels abundant. The peregrine falcon still occurs in places. Several kinds of owl are found.

Along the streams we find the dipper and the king-fisher, and the heron frequents streams and shores, though it is scarcer than it was formerly.

On the moors we hear the cry of the curlew, and grouse are abundant. Good grouse-moors are situated on the hills to the east and south-east of Lancaster. It may be noted that the red grouse is a bird confined to Britain.

On the flats by the sea are numerous birds, such as

ringed plover, turnstones, oyster-catchers, dunlins, and sandpipers.

Certain sea-fowl are scarce or absent on account of the rarity of sea-cliffs. Some of the birds are only winter visitants, but many breed in the district. Extensive breeding places are found in Walney Island and Pilling Moss.

There is little of interest with regard to the distribution of the reptiles and amphibians, but the fishes of the county present some noteworthy features. The char, a fish characteristic of the lakes of hill-regions of Britain, is found in Coniston.

Of sea-fish there is a great variety. The sandy tracts of Morecambe Bay support a great number of various species of flat-fish.

Of the invertebrate creatures much could be written in detail, but it would require considerable knowledge of zoology. Leaving the mass of these animals unnoticed, we may refer to two things. One is the abundance of edible shellfish and shrimps in the waters of Morecambe Bay; the other, the occurrence on the Coniston Fells of the mountain ringlet butterfly (*Erebia epiphron*), which, though occurring in Scotland, is not met with between the Lake District and Switzerland. Like the Alpine plants it is probably a survival from the organisms which spread over our country during the Glacial Period.

13. Climate.

The climate of a country is the result of the combined effect of the different variations of what is commonly termed the weather. The most important factors in determining the climate are temperature and rainfall.

The great variations in the climate of the world depend mainly upon differences of latitude; thus we speak of tropical, temperate, and arctic climates; that of our country being temperate. Another important factor in controlling climate over wide tracts of country is nearness to the sea ; so that along any great climatic belt we have variations according to the geographical conditions, the extremes being "continental climates" in the centres of continents far from the oceans, and "insular climates" in tracts surrounded by ocean. The continental climates are marked by great variations in the seasonal temperatures, the winters tending to be exceptionally cold and the summers exceptionally warm, whereas the climate of many insular tracts, including Britain, is characterised by equableness,—by mild winters and fairly cool summers. Again, an insular climate tends to be more humid than a continental climate. Great Britain, then, possesses a temperate insular climate.

Different parts of England possess different climates, and we must now consider wherein and why the climate of North Lancashire varies from that of other parts of the country.

Two especially important points must be regarded in

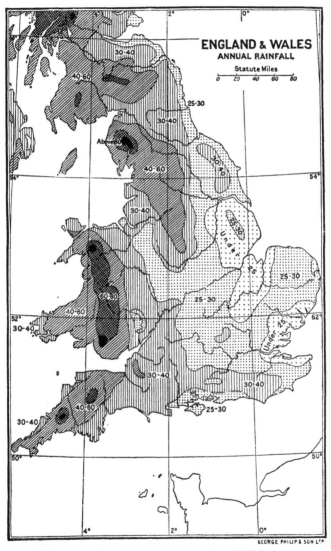

ENGLAND & WALES
ANNUAL RAINFALL
Statute Miles
0   20   40   60   80

30-40
40-60
25-30
30-40
Above 60
40-60
30-40
Under 30
Under 25
25-30
25-30
60-80
40-60
30-40
Under 25
Under 25
30-40
30-40
40-60
30-40
25-30
30-40
40-60
30-40

GEORGE PHILIP & SON LTD

(The figures give the approximate annual rainfall in inches)

contrasting the climatic conditions of North Lancashire
with those of other parts of England.   Firstly, North
Lancashire is further from the European continent and
nearer to the Atlantic Ocean than is the eastern portion
of England, and its climate therefore departs more widely
from the continental type than does that of eastern England.   In the second place the North Lancashire climate
is largely influenced by the great amount of elevated land
within the county boundaries.

As the evaporation of water and its subsequent pre-
cipitation as rain is dependent upon changes of temperature, we may consider first the temperature changes.

England and Wales are situated in a belt having a
mean annual temperature of about 50° Fahr., the mean
temperature for January being about 40° Fahr., and that
for July 60° Fahr., and these figures hold good for North
Lancashire, whereas in East Anglia the January and July
temperatures are about 38° and 62° and in parts of western
Ireland about 42° and 58° respectively.   It will be seen
then that, comparing summer and winter temperatures,
East Anglia has a less equable, and western Ireland a
more equable, climate than North Lancashire.

This distribution of temperature shows that latitude
alone does not produce the variations, otherwise it should
be colder as one passes northward.   It has long been
known that temperature variations in our island are
greatly affected by the prevalent south-westerly winds
bringing heat from the waters of the Atlantic.   These
waters off our coasts are exceptionally warm for their
latitude, owing to their movement from the warmer

south-westerly seas towards our shores on the north-east. This movement is that of the Gulf Stream, a drifting of the surface-waters of the Atlantic in a north-easterly direction caused by the prevalent winds.

It is impossible here to discuss the principles which control weather changes. It must suffice to say that our

Trees near Torrisholme bent from the south-west
(*Showing direction of prevalent winds*)

weather is largely influenced by the prevalence of *cyclones* from the Atlantic. The air movements are *cyclonic* or *anticyclonic*. In a flowing stream we may often observe a chain of eddies bounded on either side by more gently moving water. Regarding the general north-easterly moving air from the Atlantic as such a stream, a chain of eddies may be developed in a belt parallel with its general line of movement. This belt of eddies, or cyclones, as they are termed, tends to shift its position, sometimes passing over our islands, at others to the northwards and at others again to the southwards. To the shifting of this belt most of our weather changes are due.

When the country is influenced by a cyclone it is often windy, while when under the influence of an anti-cyclone it will more probably be still and dry. Cyclones, then, are apt to be accompanied by wind and rain, anti-cyclones by calm, during which there may be bright sunshine with warmth in summer, clear cold weather in winter, and fog in autumn.

There is one period of the year when the distribution of the winds in our country is affected in a different way by the temperature of the great continental mass to the east. The conditions are then such that the belt of cyclones is, as it were, pushed back over the ocean, and we experience in our county the east winds which are often prevalent during the month of March.

Let us now further consider the rainfall. Cold air can hold less water-vapour than hot air, and accordingly when the air rises and becomes chilled in the higher parts of the atmosphere it tends to part with its moisture as rain. This air may rise by expansion, which makes it lighter, or by blowing up a rising land-surface. The importance of the latter cause is great, as may be seen by studying the map of rain-distribution in our island, when it will be noticed that the areas of high rainfall coincide with the elevated regions. The prevailing winds are from the west and south, and the large amount of rain which falls in North Lancashire is mainly due to the vapour-laden winds from the Atlantic being forced up the hills and precipitating their moisture, and accordingly the greatest amount of rainfall occurs practically on the tops of the ridges which face the ocean. The greatest rainfall in the county occurs in the hilly district, in most parts of which it varies from 40 to 60 inches per annum. The heaviest fall is in the high ground of the Coniston Fells, where it amounts to about 80 inches, though much rain also descends on the hilly tracts of the south-east. The Lancashire plain between the Ribble and the Wyre receives only from 30 to 40 inches per annum. It may be remarked that the driest part of England has less than 20 inches per annum.

The amount of sunshine recorded varies in different parts of England, the greatest amount being in the south and east, and the least in the southern part of the Pennines. Along the greater part of the south-west more than 1700 hours per annum are recorded. The smallest amount for England is under 1200 hours. North Lancashire lies in a belt receiving more than 1200 and less than 1400 hours.

Severe frost is not so frequent in North Lancashire as it is in parts of south-eastern England, where the average winter temperature is lower. Snow falls in the winter season on the higher fells and often lies long there, but there is no very great amount of snow in the lower tracts.

14. People—Race, Language, Settle-ments, Population.

We have no written record of the history of our land carrying us beyond the Roman invasion in B.C. 55, but, owing to the discovery of various relics which are being revealed to us by the plough of the farmer and the spade of the antiquary, we know that Man inhabited it for ages before this date.

So remote are the times in which the forerunners of our race flourished, that antiquaries have not ventured either to date their advent, or to give an idea as to the length of time during which each division in which they have arranged them lasted. It must therefore be understood that the divisions or Ages of the times when man existed in this country before the advent of the Romans are as yet vaguely separated from one another.

When the Romans invaded our island, it was occupied by people whom we are accustomed to speak of popularly as the early Britons. These people, however, were not all of one race, and we may briefly consider who they were. Of the earliest inhabitants of our land known as the "Palaeolithic" men—men of the Old Stone Age— we have no trace save the implements which they left behind, and of these we know of none which have been found in Lancashire. It is by no means certain however that Lancashire was not inhabited by these Palaeolithic men, at any rate towards the end of Palaeolithic times.

In these times man was in a somewhat primitive state.

He did not till the land, and it is doubtful whether he had domesticated any animals. He occupied caverns, though no doubt also living in the open, and probably clothed himself exclusively in skins.

Long after the disappearance of these people, a short swarthy race arrived from the Continent and spread widely over Britain certainly penetrating into North Lancashire, as indicated by their relics. Neolithic man was in a much more advanced state of civilisation than his precursor. He tilled the land, bred stock, made rude pottery, and erected remarkable monuments. He had, nevertheless, not yet discovered the use of the metals, and his implements and weapons were still made of stone or bone, though the former were often beautifully shaped and polished. The nature of these instruments will be more fully noticed in the chapter upon Antiquities. Whence these Neolithic or New Stone Age men came and who they were we know not for certain; all we can say is that they were an earlier set of immigrants than the Celts who succeeded them.

These early men were displaced by a taller and more powerful people armed with better weapons, who, however, probably did not completely destroy their conquered enemies, but held the survivors in bondage as slaves. The more powerful race, the Celts, are supposed to have come into Britain at two distinct times. The earlier immigration was of a Celtic race who spoke a language like the modern Gaelic: these people are known as the Goidels or Gaels. Subsequently another Celtic race—the Brythons—speaking a language like Welsh arrived.

These people were acquainted with the use of metals.

Stone was gradually replaced by copper and bronze, and before the occupation of the Romans the latter had largely given place to iron.

The discovery of the method of smelting the ores of copper and tin and of mixing them was doubtless a slow affair, and the bronze weapons must have been ages in supplanting those of stone, for lack of intercommunication at that time presented enormous difficulties to the spread of knowledge. Bronze Age man, in addition to fashioning beautiful weapons and implements, made good pottery, and buried his dead in circular barrows.

There were then, in our county—even if Palaeolithic man never arrived there—three races before the Roman invasion, one pre-Celtic and two Celtic, though some believe that the inhabitants of what is now Lancashire were essentially Goidels. Be this as it may, at the time of the arrival of the Romans, the north of England, including Lancashire, was occupied by a powerful Celtic tribe, that of the Brigantes. This tribe was divided into sub-tribes, but their distribution is uncertain, and it will be sufficient for our purpose to know that the people of the Brigantes in those days inhabited what is now Lancashire. It is doubtful whether any important town of the Brigantes lay in Lancashire territory, and whether any traces of the Brigantes can be found among the characteristics of the existing people of Lancashire. But many of their place-names still survive.

In the first century A.D. the influence of the Romans began to be felt, and was exerted in Lancashire for nearly four hundred years. Important as was the civilising

influence of the Romans upon the inhabitants, as we shall
see in a later chapter when we come to speak of the roads
and other relics of that people, the Roman occupation
produced little permanent effect upon the physical
characters and the language of the inhabitants.   The
occupation was essentially military, and the Roman
legions were composed of a soldiery of mixed race
gathered by the Romans from various quarters of Europe.

After the departure of the Romans in the fifth century
we know practically nothing of what happened in the
district, except that for about two centuries it was
occupied by the Brythons or Britons.   In the seventh
century the Anglo-Saxons entered the district, and it
became part of the English kingdom of Northumbria.

In the ninth and tenth centuries Danes and Norsemen
entered the district from the north-east, and from the sea,
settling especially in the fertile tracts along the coast.

William Rufus in the year 1092 brought an army to
the north, and the Norman settlement of what is now
North Lancashire began.   This was the last important
immigration of the various races into that area.   Of these
invasions that of the Romans had a striking effect upon
the civilisation of the people ; that of the Anglo-Saxons
gave us our language (afterwards modified by Norman
influence); while that of the Danes and Norsemen doubtless considerably affected the present physical characters
of the inhabitants of North Lancashire.

It has been well said that the history of a country is
written on the face of the country itself—in the names
of its towns and villages, its rivers, mountains, and lakes.

And so we shall find that the North Lancashire place-names give much evidence of the character of the different invasions.

The effective displacement of the Britons by the Anglo-Saxons is shown by the rarity of British place-names for towns and villages, though they were still largely used for physical features.

Of British place-names we may note Leven, Ribble, and Morecambe.

Anglo-Saxon villages often end in "*ton*" (originally an enclosure) or with "*ham*." Examples are Dalton, Overton, Wennington, Aldingham, Kirkham.

Among Norse and Danish words we have "*by*" (a village), as Nateby and Kirkby; "*thwaite*" (a clearing), as Seathwaite, Allithwaite and Brackenthwaite; "*ness*" (a promontory), as Amounderness; and many others.

Traces of settlements of the pre-Roman dwellers in the district have been found in many places, often on high ground. The Anglo-Saxon, Danish, and Norse immigrants probably occupied some of the settlements which had been founded by their predecessors, but there is no doubt that they founded many new hamlets and villages.

On the arrival of the Normans, the country was parcelled out into large areas, and divided among the Norman barons, under whom the general mass of the inhabitants lived as bondsmen, but no doubt those who dwelt in the more remote uplands remained undisturbed in the possession of their small freehold estates.

The population of Lancashire as a whole, as stated in

Chapter 2, was 4,768,474 in 1911, that of North Lancashire being about 400,000. With the exception of the
county of London it has the largest population of all the
English counties. It has an average of 2550 people per
square mile as compared with the average of 618 per
square mile for the whole of England and Wales. North
Lancashire, however, where the population is much less
dense, has only about 640 people per square mile.

15. Agriculture. Forestry.

As many of the inhabitants of North Lancashire have
from time immemorial essentially subsisted by agriculture,
we must devote some attention to it.

In doing so, it is well to note first the principal factors
which control the agricultural operations of any given
county or district. These are latitude, altitude, climate,
soil, character of the people, and markets for the produce.

Latitude is of special importance in affecting the
nature of the corn crops, for the county is too far north
to allow of successful cultivation of wheat on a large
scale, while on the contrary it suits the growth of oats,
and accordingly we find that this is the chief kind of corn
grown.

Altitude is responsible for the dividing line between
the areas devoted to the cultivation of crops and to
permanent pasture. The area of the former is chiefly in
the lowlands and valley-bottoms, whereas much of the
latter is on the upper portions of the fells. The line
separating the tracts where arable cultivation is profitable

from those of permanent pasture may in North Lancashire be roughly placed at a height of 900 feet above sea-level.

It will have been gathered from the remarks made in a previous chapter that the county enjoys an equable climate with fairly warm winters and cool summers, and that it is essentially humid. Such a climate is suitable to the growth of many crops, and has always favoured grass rather than arable farming.

We have considered the soil in the geological chapter. It was there seen that there were four important varieties, namely the clayey soil of the slate-rocks, the thin light soil of the Mountain Limestone tracts, the sandy soil of the Millstone Grit and New Red Sandstone regions, and the mixture of peat and silt bordering parts of the coast and occupying the sites of infilled lakes. The agriculture of each of these soils is marked by special features.

The character of the inhabitants is also of great importance. The physical strength, capacity for work, and tenacity of purpose largely inherited from their Scandinavian ancestors and fostered by the struggle against physical conditions which cause the produce to be wrung from the earth by hard labour, have enabled the natives to overcome the difficulties with which they were confronted, when a weaker and more indolent race would have been worsted in the struggle.

Lastly, there is the market for the produce. In early days corn and roots were grown in sufficient quantity for home consumption only, and the same may be said of the supply of meat. The wool of the sheep alone was carried to the markets to exchange for such necessaries of life as could not be obtained on the spot. Of recent years the growth of large towns, including watering-places, in the northern part of the county has taken place to a degree sufficient to ensure the consumption of agricultural produce within the district, and little is sent further afield.

The following figures from the *Journal of the Royal Agricultural Society for* 1910 give the acreage devoted to the different corn-crops and the number of live-stock in the whole of Lancashire (see pp. 179, 180). Figures for the northern part of the county are not available, but the proportions of the different crops and live-stock given below are roughly applicable to the special district under consideration :—

Corn Crops					*Acreage*
Wheat	...	...	...	...	19,757
Barley	...	...	...	...	4,427
Oats ...	...	...	...	...	83,786
Live Stock					*Numbers*
Cattle ...	...	...	...	...	227,416
Sheep ...	...	...	...	...	324,484
Pigs ...	...	...	...	...	69,932

Much of the sheep-pasture, as already stated, is on the higher hills, which are largely occupied by the old slate rocks in the part north-west of Morecambe Bay and by the Millstone Grit in the area south-east of Lancaster, though parts of the Mountain Limestone, with its sweet herbage, form excellent pasture.

In the district north-west of Morecambe Bay, many of the sheep are of the Herdwick breed, of which tradition says that they escaped from a vessel of the Spanish

Armada wrecked on the west coast of Cumberland, and thence gradually spread through the district.

On the borderland between Lancashire and York-shire, where the formation is Millstone Grit, is a very characteristic breed of sheep known as "Lonks" (i.e. Lancashire sheep, which is often pronounced "Lonki-

"Lonks" Sheep on Millstone Grit Fells

shire"). They are found on the fells south-east of Lancaster about Roeburn Dale and the Trough of Bowland.

The district is not one of pig-breeders. Those pigs which are used for the consumption of the offal of the towns are brought from elsewhere.

The cattle consist almost entirely of the Large Dairy Shorthorns. These are used during the early part of their lives in making the famous Lancashire cheese. They are afterwards absorbed into the cow-houses of the intensive suburban farmers, who supply milk from them, and feed them out fat at the same time. These farmers are very characteristic of the districts around the larger centres of population. Everything produced on the farm is sold off; their produce includes hay, straw, potatoes, and milk. A high state of fertility is nevertheless kept up by large purchases of manure and concentrated feeding-stuff.

One of the most interesting features of the district, from the agricultural point of view, is the great contrast, in a small tract, between the highly developed intensive farming of the lowlands around the large centres of population and the primitive farming of the tracts where the Lonks sheep are reared.

In the Fylde district very large city dray-horses of the Shire type are bred, and are mostly sent to Liverpool.

In the lowlands a certain amount of arable land exists, especially within easy access of the larger towns. From the figures given it will be seen that the growth of wheat and barley almost entirely gives place to that of oats, and this is especially true of North Lancashire. Practically the whole of the wheat is grown for its straw. Of root-crops, we find in the intensive farm districts, in addition to the potatoes already mentioned, crops of mangolds and swedes, and a considerable amount of rough garden-produce.

Forestry is not very important in North Lancashire. In addition to trees grown for their timber are others for ornamental purposes, for game-covert, and for shelter-belts around hillside farm-houses. For the latter purpose the sycamore is an ideal tree, and one which thrives well here.

Lakeside: Bobbin Mill

The comparative lack of trees is partly due to poor soil, and partly to exposure. Shelter is of more importance than soil alone, and we find trees growing profusely here and there in the cloughs and sheltered valleys. Even the Millstone Grit area, when covered with mild humus in the lower valleys and lower hillsides, grows very fine crops of deciduous trees and many conifers.

The lowlands are comparatively treeless. Not only are these lowlands exposed, but much sea-spray is very injurious, especially to evergreen plants when carried inland during gales.

Climatic changes have certainly caused the diminution of tree-clad areas. In many of the upland peaty tracts remains of former forests are seen, and the mild humus of these former forests has been replaced by an acid humus which is decidedly unfavourable to tree-growth. In former days the forests of Quernmore and Upper Wyresdale were of importance.

16. Industries and Manufactures.

Before treating of the more important industries, notice may be taken of a few minor ones which are extinct or dying out, but are nevertheless of historic interest. They belong chiefly to the district north of the sands.

The working of iron-ore will be noticed in the next chapter. The old "bloomeries" or forges where iron was smelted will also be noticed in that chapter.

The abundance of coppices in Furness and Cartmel especially is responsible for the utilisation of the wood for the manufacture of bobbins, baskets, and hoops, and charcoal burning is still carried on in those districts in connection with the manufacture of gunpowder.

The coarse Millstone Grit of the fells east of Lancaster furnished material for the manufacture of handmills or querns. Quernmore, near Lancaster, takes its name from this now extinct industry. The manufacture of silk has greatly declined in the county.

At the present time the two chief industrial towns of North Lancashire are Preston and Barrow-in-Furness. Preston is a centre of the cotton industry. According to the census of 1901 there were 24,950 people in that town

Charcoal Burners in Cartmel District

engaged upon it. The cotton trade of Lancashire is so important that we must say more about it, though apart from Preston it belongs to the southern part of the county. The manufacture of cotton is facilitated by the damp climate of the county and has been carried on since the sixteenth century. The eighteenth century was marked by the Industrial Revolution, of which a most important cause was the introduction of steam-power for driving machinery; and the factory system was substituted for the system of working in small houses.

The two processes of spinning and weaving, which were more than once united, have now become separated for a variety of reasons, and while spinning is now extensively carried on in South Lancashire, weaving is mainly conducted in the towns immediately to the south of the Ribble, and in Preston.   According to the census returns of 1901 there were only 2,036,000 spindles employed in Preston as against 57,900,000 looms.

To give an idea of the importance of Lancashire from the point of view of the cotton-trade it may be mentioned that at the beginning of this century about 400,000 out of a total of nearly 500,000 cotton operatives employed in England and Wales worked in our county.

Engineering and machine-making are also carried on at Preston, and according to the same census employed 3016 people.

Barrow-in-Furness is especially connected with the steel-trade.   The small forges scattered about the Furness district disappeared as the result of improvements in the processes of smelting which came into use about the end of the eighteenth century ; and the Bessemer process of steel manufacture discovered in 1856 led to the formation of the great works at Barrow.

The 1901 census showed that 1765 people were engaged in the iron and steel manufacture.   In the same town 5496 people were engaged in engineering and machine-making.   The introduction of iron ship-building

Bobbin Mill, Hawkshead

in 1870 was a very important step as regards the prosperity of Barrow, and this industry now employs 10,300 men. The manufacture of guns is also an important industry in Barrow.

Steel is also manufactured at Askham, Ulverston, and Carnforth.

Several industries are connected with the town of Lancaster, the most important being the manufacture of linoleum and oilcloth.

The taking of fish is one of the great industries of the British Isles, but the fishing stations on the North Sea are much more important than those of the western coasts. Nevertheless, the capture of fish (especially if we include shellfish) around the coast of Morecambe Bay is very considerable. About 366 boats and 1278 men are engaged in the fisheries of the North Lancashire coast.

On the north side of the bay the principal fishing ports are Barrow, Ulverston, and Cark, but these are not of great importance except as regards cockles and shrimps. Morecambe is noted for its mussels, shrimps, and flat-fish. The best mussel beds lie between Morecambe and Heysham, on the tracts of stone or "skears" which there abound. The so-called prawns of Morecambe Bay are not true prawns, and may be passed by with mere mention. The supply of cockles, mussels, and shrimps on the other hand is very large. Cockles are obtained nearly all the year round, but chiefly in winter and spring. They are secured by means of a three-pronged iron fork attached to a wooden haft: this is known as a "cockle-craam."

Furnace, Ulverston Ironworks

The mussels are obtained by means of a rake attached to a shaft thirty or forty feet in length: this is a "long craam."

Fleetwood is the greatest of the Lancashire fishing-ports, and is indeed one of the most important in England.

Fleetwood Fishing Boats

There is much shrimp-fishing at Lytham and St Anne's-on-Sea. Salmon are extensively caught in the larger rivers.

17. Mines and Quarries.

The chief mines of North Lancashire are those which yield iron : they are situated in the Furness district in the area composed of Carboniferous Limestone around Dalton and Lindale. The ore is an oxide of iron known as red haematite, occurring chiefly in fissures of the limestone.

When these deposits were first worked is unknown, but scattered about along the lake sides and stream sides of High Furness are the remains of old smelting-works or *bloomeries*, as they are termed, where the ore was reduced by means of charcoal. These bloomeries were erected where charcoal could be readily obtained, and in many cases also where water-power was available. They occur on Peel Island, in Coniston Lake, in Rusland, on the shore of Windermere, and in two or three places in the Crake valley. The old bloomery of Duddon Bridge still stands just outside the county, on the Cumbrian side of the Duddon. This was in use at the end of the eighteenth century. The furnace at Backbarrow below the foot of Windermere is still at work. At Peel Island medieval remains have been found associated with the bloomery.

In the latter part of the thirteenth century ore was certainly worked in connection with Furness Abbey, but the extraction and smelting of the ore on a large scale are comparatively recent.

The *General Report and Statistics* published by the Home Office in 1908, which contains the latest official figures, states that the Furness district supplied 394,843 tons of iron ore in 1907, which ore contained 50 per cent. of available iron. The total value at the mines was £307,273, being 15s. 6¾d. per ton.

The ore was derived from the following localities :—

				Tons
Askham and Dalton	...	...	...	15,294
Elliscales	...	...	...	8,509
Lindal Moor	...	...	...	56,154
Newton	...	...	...	1,500
Park ...	...	...	...	119,373
Roanhead	...	...	...	194,013
		Total	...	394,843

Other ores which have been somewhat extensively worked in past times are those of lead and copper, though at the present day various causes have led to the decline of the mining of both of these metals in the county. The lead ore, chiefly galena or lead sulphide, occurs in the Coniston Fells, where however copper mining was much more important. The working of the copper mines certainly began here in very remote times. The veins of these ores occur in the volcanic series of the Ordovician rocks. Copper ore was also worked in past times in the Carboniferous Limestone rocks of Warton Crag.

A thin seam of coal occurs in the Millstone Grit series east of Lancaster, and was once worked locally.

Many good building-stones are found in the northern part of the county. The rocks of the volcanic series are

Slate Quarry: Tilberthwaite

used locally for building-stones in the neighbourhood of Coniston, and those of the Silurian rocks in High Furness. The Carboniferous Limestone also is extensively used in those tracts where it occurs, and Millstone Grit is largely quarried in the southern part of the area. The town of Lancaster is mainly built of this stone. The New Red Sandstone is quarried near Barrow. Furness Abbey is built of it.

Slate of a green colour is largely worked in a seam of the volcanic series which runs from Little Langdale to the Duddon. The principal quarries are in Tilberthwaite, on Coniston Old Man, and at Walney Scar. Another band of slate of the higher Ordovician rocks is worked on Broughton Moor, and a more important band in the Silurian rocks is quarried on the same moor, and very extensively in the Kirkby quarries, east of the Duddon estuary.

Flagstones are obtained from some of the Silurian rocks of Furness, and also from some flaggy beds of the Millstone Grit of the fells east and south-east of Lancaster. Clay for making bricks is obtained from some of the Millstone Grit shales.

Lime is burnt in many places where the Carboniferous Limestone is developed, and is used for various purposes.

Rock-salt is extracted from the New Red Sandstone at Preesall, and is used in the United Alkali Company's works at Fleetwood and elsewhere. The output is about 3000 tons per week.

Finally peat for fuel is dug from the turbaries or peat

Peat Diggers: Cark-in-Cartmel

bogs on the higher hills in several places, and also on the flat mosses which are found on the low grounds bordering the estuaries and Morecambe Bay.

18. History of North Lancashire.

The Roman occupation of Britain began with the visits of Julius Caesar in B.C. 55 and 54, but the Romans first arrived into what is now Lancashire territory under Agricola in A.D. 79. That event marks the beginning of the history of Lancashire, earlier events being regarded as belonging to the prehistoric period. For nearly four centuries the Romans dominated the district, with occasional raids by the Picts and Scots, and an insurrection of the Brigantes which was soon quelled.

The Romans arrived from the south along the Lancashire plain, and probably after crossing the sands of Morecambe Bay, marched along the flat tracts of Cartmel and Furness by the sea. They opened up the country by constructing an elaborate series of roads which will be noticed subsequently, and they protected these by military camps at points of strategic importance. Under Agricola especially the Brigantes were introduced to the civilisation and luxury of the Roman conquerors.

In addition to the military posts many small towns and settlements arose in North Lancashire. In the fourth century the decline of the Roman empire had begun, and at the beginning of the following century the Emperor Honorius gave to the Britons their independence.

Shortly after the departure of the Romans, who were gradually leaving the country between 410 and 430, the English invasion began by the arrival of Teutonic peoples from the country near the Elbe. There were three Teutonic tribes dwelling near the Elbe, namely the Saxons around the mouth of that river, the Angles further north, and still further north the Jutes. All these people were termed English, and as they conquered portions of our country it in turn became English. By the end of the sixth century they had annexed the lowlands of England, the Jutes occupying Kent, the Saxons a tract north and west of Kent, and the Angles the centre, east, and north-east of England.

All this time the North Lancashire area remained part of an independent British kingdom extending from the Dee to the Clyde, and bounded on the east by the Pennine Hills. The kingdom was called Strathclyde and was divided into a series of smaller States, one or more of which included what is now North Lancashire. We know little of what went on in this region between the departure of the Romans and the arrival of the English, Danes, and Norsemen, but there were in all probability dissensions among the different States.

Some time after the battle of Chester in 607, when Ethelfrith, King of Northumbria, defeated the British, North Lancashire became English.

In the ninth and tenth centuries invasions of the Danes and Norse from the north-east and from over the sea took place, and these Northmen occupied considerable tracts especially in Lancashire north of the sands and in Amounderness, the plain north of the Ribble estuary.

Obscure as are the events which happened between the departure of the Romans and the arrival of the Normans, their influence upon the inhabitants of the area was most marked. The people of the Brigantes had disappeared, the inhabitants had largely acquired their present characteristics, and the English language had become established.

About the time of the Norman conquest the North Lancashire district was laid waste. This may have been partly due to a struggle between Harold and Tostig in 1066, but also to William the Conqueror's devastation of the Northumbrian kingdom in 1069.

About 1072 South Lancashire, then known as the land between Mersey and Ribble, with parts north of the latter river, was given by the Conqueror to Roger of Poitou, and William II granted more land north of the Ribble to him before 1094. Roger then held practically the whole of what is now Lancashire.

In 1102 Roger forfeited this land to the crown, and Henry I granted it to Stephen. After Stephen came to the throne, David King of Scotland became possessed of Lancashire north of the Ribble, but it once again and finally became part of England in the reign of Henry II, when, as we have already stated, "the County of Lancaster" was definitely established.

Little of importance took place in the county from this time until the fourteenth century. A Scottish raid in 1316 was unimportant, but in 1322 another raid was conducted with Bruce as leader, when the town and castle of Lancaster and the town of Preston were burnt.

In 1349–50, our county, like the rest of England, suffered heavily from the Black Death.

In 1351 Edward III conferred upon his cousin Henry the title Duke of Lancaster, and the Duchy of Lancaster therefore came into existence. He also conferred upon him the rights of a palatine earl. This made the county a county palatine—a little kingdom, in a sense, within the English kingdom.

After a lapse, the dukedom was restored in 1362 and John of Gaunt, fourth son of Edward III, was created Duke; in 1377 he also obtained the palatine rights. The county is still a County Palatine and the reigning monarch Duke of Lancaster, though it must be remembered that the Duchy of Lancaster and the County of Lancashire are not the same thing.

During the Wars of the Roses in the fifteenth century, the northern part of the county was not the scene of any important occurrence, but the contrary was the case during the civil war in the time of Charles I. In 1643 the parliamentarian forces took Preston and Lancaster, which were later retaken by the royalists, and in 1648 the Scottish army entered England and was defeated by Cromwell at Preston, and subsequently in South Lancashire at Wigan and Warrington, after which the victorious army returned to London and demanded the punishment of the king, who was beheaded in the following year.

North Lancashire was visited by the Pretender's

troops in 1715 when after some skirmishes near Preston they surrendered to the king's forces, and in 1745 the troops of Charles Edward Stuart went southwards through Lancashire, and retraced their steps northward during their precipitate retreat.

Since then the history of the county is chiefly concerned with its industrial progress.

19. Antiquities—Prehistoric,   Roman, Anglo=Saxon,   Norse,   Medieval.

Our knowledge of the history of the inhabitants of Lancashire is chiefly derived from a study of written records, but it is not entirely dependent upon it, for relics of the early inhabitants afford information even of those periods of which records were made in writing.

The part now forming the county of Lancashire was however inhabited in times earlier than those concerning which we have the first written records, and our knowledge of the state of the inhabitants during those early days is derived solely from an examination of relics left by them, either structures such as grave-mounds and stone-circles, or various weapons and other articles which have resisted decay and been preserved to the present day.

The periods previous to those of the first written records are usually spoken of as "Prehistoric," and we will now consider the nature of the remains which have come down to us from these times.

Before the use of metal for forming tools and weapons was discovered these were chiefly of stone, and we are

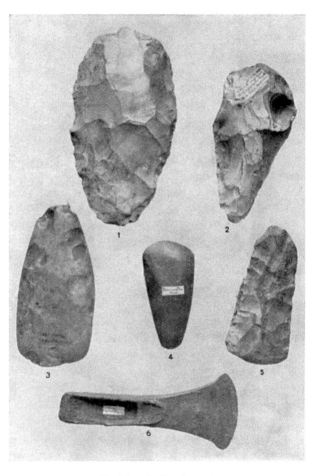

Prehistoric Implements

1, 2, *Palaeolithic*; 3, 4, 5 *Neolithic*; 6 *Bronze Age*

enabled therefore to divide prehistoric time into the Stone Age and the Prehistoric Metal Age. We will begin with the earlier of these ages—that of stone.

Antiquaries have found that there are two very different classes of stone implements marking two quite distinct ages of civilisation, of which the later was far more advanced than the earlier. We speak therefore of the Palaeolithic or Old Stone Age, and the Neolithic or New Stone Age. In the older age the implements were formed by chipping the stone into shape, and the art of grinding and polishing them was unknown. Instruments of this type are found in the river-gravels and caverns of England as far north as Derbyshire, but are unknown in the northerly parts including Lancashire. We need not, therefore, dwell further upon the remains of the older Stone Age.

Several implements of the later or Neolithic Age have been found in the northern part of the county. Among them are "celts"—stones which have been chipped and ground into the form of a broad chisel with a sharp cutting edge at the broader end. They are often polished and were probably used as hatchets. Other types are perforated hammers and hammer-axes, which have been found in various parts of Furness and at Claughton near Garstang. It is not to be supposed that all the stone implements found in this region necessarily belong to the Stone Age, for stone was used long after the introduction of metal, and has indeed been in use quite recently for some purposes, as for instance the "strike-a-light" used for igniting tinder.

There are few structures in the county which were

of certainty made by men of the Stone Age, for there is much difficulty in distinguishing the relics of this age from that of the succeeding period. "Barrows," or burial-mounds, often largely made of stones, exist. Of these barrows there are two types, the long and the round barrows. The long barrows belong to the early Neolithic period and the round barrows to a later period. Where stone is abundant cairns were often erected instead of barrows.

From study of the relics of the later Stone Age found in other places we know that the English dwellers of this

Bronze Spearhead found at Winmarleigh

age were hunters and fishermen, possessing domestic animals, and having some knowledge of agriculture. They were also acquainted with the art of making rude pottery.

The introduction of metal was certainly gradual, and long after objects formed of metal were introduced stone no doubt continued to be the principal material from which implements were fashioned. The first metal to be used for the purpose was not iron, but the alloy of copper and tin which we call bronze, for the art of smelting iron was much more difficult than that of making bronze. Accordingly the earlier prehistoric age of metal was a

Stone Circle on Birkrigg Common

(*West side of Cartmel Sands*)

Bronze Age. At first the Bronze Age men imitated the stone implements which were in use, and a large number of the bronze instruments are more or less modified forms of the hatchet-shaped stone instrument, but the introduction of metal allowed of the formation of a far greater variety of forms than could be fashioned in stone, and in our county, besides the hatchet-shaped forms, bronze dagger-like weapons and spears have been discovered. A horde of these bronze weapons was found at Winmarleigh.

Some of the men of the Bronze Age dwelt in caves, and remains of this period have been found in Kirkhead Cave near Cartmel.

The county is rich in prehistoric structures besides the barrows already mentioned. Some of these structures may belong to the Neolithic period, while others are certainly of the Bronze Age. Of the so-called "Druidical Circles," which have, of course, nothing to do with the Druids, we may mention a double circle on Birkrigg Common, east of Great Urswick.

Remains of settlements of prehistoric tribes are also frequent. Various camps and earthworks of pre-Roman date also occur, one of the most remarkable being that on the summit of Warton Crag: another is on Pennington Hill near Ulverston. Circles, cairns, and the remains of settlements are found on the slopes of the Coniston Hills, on Birkrigg Common, and at Urswick near Ulverston.

There is evidence that iron had been introduced into Britain before the arrival of the Romans into our island, and a few relics found in North Lancashire have been assigned to this Early Iron Age.

We may now proceed to consider the relics of the Roman occupation of our land. In so doing we pass definitely from prehistoric to historic times. Of the Roman roads we shall speak elsewhere, and at the same time refer to the most important of the Roman camps.

The chief Roman relics of small size consist of pottery of very artistic types, some of which was made in Britain, though a large part was imported from the Continent, especially from Gaul. There are also many and various ornaments, and a large number of Roman coins, which have been found in several places. Many inscribed stones have been discovered; a Roman milestone was found at Caton, and altars at Ribchester.

The Roman sites at Overborough, Lancaster, and Ribchester, to be noticed more fully in a subsequent chapter, have yielded a very large number of Roman remains.

There are many relics belonging to the period between the departure of the Romans and the arrival of the Normans, the most interesting of which are sculptured stones. Some of these are of Anglo-Saxon origin, while others show Norse influence. Of Anglo-Saxon relics we may notice a silver cup and other objects found at Halton, and some bronze brooches unearthed at Claughton.

Many pre-Norman crosses exist. Several have been found at Lancaster, others at Bolton-le-Sands, Hornby, Heysham, and Halton. Some of these show Norse work. One at Halton is of interest as marking, according to some authorities, the replacing of pagan worship by the Christian religion. The curious procumbent tombstones

The Hogback, Heysham Churchyard

(Showing the two sides)

known as hogbacks belong to this period. One was found at Bolton-le-Sands, and a beautiful specimen exists in that remarkable collection of pre-Norman relics—the churchyard at Heysham.

Of possible Norman age are certain mounds usually surrounded by ditches, and often having courts or baileys attached. They are usually on low ground near rivers or the sea. In the Lune valley they are found at

The Lancaster Penny

Whittington, Arkholme, Melling, Hornby, Halton, and Lancaster, and on the Ribble at Preston. A very large example at Aldingham is on the coast.

Medieval remains other than buildings are of no great interest. Examples of the rough pottery characteristic of the period are frequently unearthed.

Coins and tokens seem to have been struck at Lancaster at various times between the reign of Henry II and the end of the eighteenth century. The illustration shows the last of these tokens, known as the Lancaster Penny.

20. Architecture—(a) General.

It will be convenient to consider under these headings the ecclesiastical, military, and domestic buildings of the county. Before doing so, however, we may offer some observations on the edifices in general.

We may remark at the outset that here as elsewhere the buildings are affected by the nature of the materials available, local stone being mainly used. Thus in the area of the Ordovician rocks, stones of that age are chiefly seen; in the Carboniferous tracts, the limestones and sandstones belonging to that geological system have been employed. These Carboniferous rocks, and especially the sandstones, form the main building materials in our area.

A preliminary word on the various styles of English architecture is necessary before we consider the churches and other important buildings of our county.

Pre-Norman or, as it is usually, though with no great certainty termed, Saxon building in England, was the work of early craftsmen with an imperfect knowledge of stone construction, who commonly used rough rubble walls, no buttresses, small semi-circular or triangular arches, and square towers with what is termed "long-and -short work" at the quoins or corners. It survives almost solely in portions of small churches.

The Norman Conquest started a widespread building of massive churches and castles in the continental style called Romanesque, which in England has got the name of "Norman." They had walls of great thickness, semicircular vaults, round-headed doors and windows, and massive square towers.

From 1150 to 1200 the building became lighter, the arches pointed, and there was perfected the science of vaulting, by which the weight is brought upon piers and buttresses. This method of building, the "Gothic," originated from the endeavour to cover the widest and loftiest areas with the greatest economy of stone. The first English Gothic, called "Early English," from about 1180 to 1250, is characterised by slender piers (commonly of marble), lofty pointed vaults, and long, narrow, lancet-headed windows. After 1250 the windows became broader, divided up, and ornamented by patterns of tracery, while in the vault the ribs were multiplied. The greatest elegance of English Gothic was reached from 1260 to 1290, at which date English sculpture was at its highest, and art in painting, coloured glass making, and general craftsmanship at its zenith.

After 1300 the structure of stone buildings began to be overlaid with ornament, the window tracery and vault ribs were of intricate patterns, the pinnacles and spires loaded with crocket and ornament. This later style is known as "Decorated," and came to an end with the Black Death, which stopped all building for a time.

With the changed conditions of life the type of building changed. With curious uniformity and quickness

Furness Abbey: Transitional Arches

the style called "Perpendicular"—which is unknown abroad—developed after 1360 in all parts of England and lasted with scarcely any change up to 1520. As its name implies, it is characterised by the perpendicular arrangement of the tracery and panels on walls and in windows, and it is also distinguished by the flattened arches and the square arrangement of the mouldings over them, by the

St Patrick's Chapel, Heysham

elaborate vault traceries (especially fan-vaulting), and by the use of flat roofs and towers without spires.

The medieval styles in England ended with the dissolution of the monasteries (1530–1540), for the Reformation checked the building of churches. There succeeded the building of manor-houses, in which the style called "Tudor" arose—distinguished by flat-headed windows, level ceilings, and panelled rooms. The ornaments of classic style were introduced under the influences of Renaissance sculpture and distinguish the "Jacobean" style, so called after James I. About this time the professional architect arose. Hitherto, building had been entirely in the hands of the builder and the craftsman.

There is one character which is common to many of the buildings of the county, whether ecclesiastical, military, or domestic, and as it is one of particular geographical import, it must be noticed. North Lancashire, not far from the borders, has witnessed much fighting and has been the scene of raids, consequently other edifices than those erected exclusively for military purposes required fortifying. The quadrangular tower, the nucleus of the Norman fortification, served as the pattern for other fortified buildings, and accordingly the tower of this type forms a noteworthy feature of many ecclesiastical and domestic buildings.

21.   Architecture — (*b*)   Ecclesiastical. Churches and Religious Houses.

The ecclesiastical buildings of North Lancashire include the churches and the religious houses.

Many of the churches have undergone much restoration, and although in some cases the ancient character has been preserved during the process, in too many it has been completely destroyed. We may give examples of various ecclesiastical buildings which illustrate the different styles of architecture.

Of pre-Norman buildings we have no large examples

such as are found in other parts of the country, but
certain very interesting remains are found in the promon-

Norman Doorway, Overton Church

tory which lies south of Morecambe between the bay
and the estuary of the Lune.   In this promontory, far
from the lines of main traffic until quite recently, we

may well expect that buildings, which if situated elsewhere would have been destroyed, would escape the hands of the spoiler.

Parts of the fabric of St Peter's Church, Heysham, has been claimed as pre-Norman. It is marked by thick walls, with hard mortar, and stones built in somewhat

Early English Doorway of Norman type,
Gressingham Church

irregularly. The church contains a narrow round-headed doorway. The most interesting relic of the times of which we speak is, however, the little chapel of St Patrick which stands on the cliff near St Peter's Church. It was but 24 feet long and 8 feet wide; only the west walls and part of the south side are now standing. In the latter are remains of a window, and a narrow doorway, surmounted by a semi-circular arch formed of a single stone ornamented with semi-circular flutings.

The Norman period is represented by some interior work, and several doorways. In Cartmel Priory, the pillars of the interior supporting the tower are of this date, as are also the arches inside Hawkshead Church, and those of the south aisle of Aldingham Church. Doorways of Norman architecture may be seen in the churches of Ulverston, Broughton-in-Furness, Kirkby Ireleth, and Overton.

Examples of the style of the period of transition from Norman to Gothic are found in the small church of Styd near Ribchester, which is 18 yards long and 9 yards broad, and in the church of Furness Abbey, which was chiefly built at this time, though it underwent many subsequent alterations. The doorway of the north transept of that church illustrates this period of transition.

The Early English period is represented by some of the buildings of Furness Abbey: the Chapter House furnishes admirable examples. The windows in the north transept of Cartmel Priory, the doorway at Gressingham, and the south door of Styd Church may be quoted as examples of this period.

Various minor examples of the Decorated period are found. Some beautiful Decorated windows have unfortunately been removed from Urswick Church, but are preserved in a garden at Hawkshead. The infirmary of Furness Abbey is of this period, but some of the best representatives of Decorated architecture are found in

Cartmel Priory. Special mention may be made of the beautiful windows in the south aisle of the chancel, as good examples of a special style of late Decorated work.

Much Perpendicular work is found throughout our district. A great part of Lancaster parish church is in this style. Better examples are various windows in the

Cartmel Priory Church

(*Decorated windows on right, Perpendicular windows in middle*)

church of Furness Abbey, much of Aldingham and Warton churches, the towers of Hornby and Bolton-le-Sands, and windows in the transepts, nave, and Piper Choir of Cartmel Priory.

Let us turn now to the Religious Houses, some of

which have been mentioned to illustrate styles of architecture.

Of Abbeys, the principal was the great establishment of Furness, whose monastic buildings are situated in a secluded valley not far from the southern end of the Furness peninsula. It was founded by monks of the Benedictine Order in 1127, but in 1148 became a Cistercian abbey. The Abbot was a person of great importance, and was chief lord of the liberties and royalties of Furness. In addition to most of the lands of Furness, the Abbey erected and possessed many buildings, as the manor-house at Hawkshead Hall, and the castle at Piel. It had much cultivated land, forests, hunting-grounds and fisheries. The Abbey was surrendered to Henry VIII in 1537. Among the existing portions of the Abbey are remains of the church; and of other conventual buildings the chapter-house, fratry, infirmary, and part of the Abbot's house still exist.

Cockersand Abbey, situated just south of the Lune estuary, was founded at the end of the twelfth century, on the site of a pre-existing hospital, as a House of the Premonstratensian monks. Of the buildings, the octagonal chapter-house alone remains.

The Cistercian Abbey of Wyresdale was an offshoot of Furness Abbey, and was founded towards the end of the twelfth century. Its actual site is unknown.

Priories existed at Lancaster, Cartmel, Conishead, Hornby, and Cockerham.

St Mary's Priory, Lancaster, was the first religious house in Lancashire. It was founded by Roger of Poitou

Cartmel Priory Church : Interior

in the reign of William II. The monks were of the Premonstratensian Order. No remains of this priory now exist.

Cartmel Priory was founded in 1188 by the Earl of Pembroke as a priory of Augustinian canons, but a church stood here before the foundation of the priory. Reference has already been made to the existing Priory Church.

Conishead Priory, on the west side of the Leven estuary, was founded by Gabriel de Pennington in the reign of Henry II as a hospital for lepers and poor persons, but was soon converted into a priory of the Augustinian monks. No remains of the priory now stand, and the present building of that name is a modern erection.

Hornby Priory was a cell belonging to the Premonstratensian abbey of Croxton in Leicestershire. A few sculptured stones are still found in the grounds of the Priory Farm.

The Augustinian Priory of Cockerham was established in 1207 or 1208. No remains of the building are left standing.

The site of a Franciscan friary stands about half a mile from Preston market-place, and that of a Benedictine monastery at Tulketh, near Preston, on the north side of the Ribble.

22.   Architecture—(c) Military and other Castles.

The division into military, ecclesiastical, and domestic architecture is, for North Lancashire, somewhat arbitrary. Several of the castles were not strictly speaking military, though designed for defence; but this was also the case with many of the manor-houses, and even ecclesiastical buildings in the troublous borderland were to some extent fortified.

Lancaster Castle was in a strict sense a military castle, and it is by far the most important of the castles of North Lancashire.

In the first place let us consider the natural advantages of its site. The Normans were quick to seize upon sites which possessed such advantages, though, as in the present case, they frequently chose places which the Romans, with their genius for the selection of strategic positions, had previously occupied.

Standing on the high ground on which the castle is built one looks eastward over the high moorlands which stretch away towards the Pennines, and to the west are the flats of the Lancashire plain which at no distant date were a network of mosses and marshlands. Between highland and marshland the great route to the north lay, and crossed the Lune over the shallows. Overlooking these shallows is the northern end of a ridge which lies between a little valley, in which most of the modern town of Lancaster stands, and the low ground of the Lune to the north-west.   Upon this northern end the castle was erected.

The Norman castles as first built consisted of quad-

Lancaster Castle : Entrance Gateway

rangular towers or keeps.   Subsequent additions were made to the castles in the form of a wall with subsidiary towers surrounding the ward or inner space on which stood the keep, while other subsidiary buildings might be erected in this ward.

The castle was apparently built so that about half lay within and half without the old Roman wall. The keep was erected by Roger of Poitou before 1102, and it is the only part of the castle which has retained anything like its original appearance.

Of the present buildings, much of the surrounding wall with its towers is modern, but parts of older buildings still stand. The great entrance gateway was probably built early in the fifteenth century. It contains windows of Perpendicular style of architecture. To the south of the keep is a tower known popularly as Adrian's tower. It has a modern exterior, but inside it is some work showing the style of the period transitional between the Norman and the Early English styles, and probably dating from before 1200. The old dungeon tower is now destroyed, but the interior of the Well Tower north of the gateway has some transitional work. The upper part of the keep was rebuilt in the reign of Queen Elizabeth. Long after Norman times the castle was of importance in connection with the frequent raids and guerrilla wars conducted by the Scotch borderers. It was ultimately used as a prison, but even in this state important events are connected with it, including the imprisonment of the "Lancashire witches" before their execution in 1612, and of George Fox the founder of the Society of Friends in 1664.

Piel Castle, or the Pile of Fouldrey, situated on an island at the southern extremity of the Furness peninsula, was built by the monks of Furness Abbey in Stephen's reign, as a protection against raiders from Scotland. It was repaired and strengthened in the fourteenth century. The existing portions are part of the keep and remains of inner and outer walls, which once enclosed the keep. Four towers of the inner walls stand, one of which is a gate-tower, and there are three towers of the outer wall. Lambert Simnel landed here in 1487.

Piel Castle

The other castles were built by members of important Lancashire families. Gleaston Castle, near Aldingham, in the Furness peninsula, is said to have been erected by the Harringtons. The date of erection is unknown, but its style points to the early part of the fourteenth century. Little now remains.

Hornby Castle, on a hill overlooking the Lune and Wenning, about nine miles above Lancaster, is first mentioned in 1362. The chief event connected with it occurred when it belonged to the Stanleys—the attack by the Scotch in 1513. Little of the old edifice is left. Thurland Castle, the seat of the Tunstalls, is a few miles higher up the Lune. The original building was

Wraysholme : showing Peel-tower

erected in the reign of Henry IV, but the existing edifice is almost wholly modern.

Greenhalgh Castle, close to Garstang, was built by the Earl of Derby in 1490 for the protection of his estates. It was held for the king by a later earl in 1643, and was afterwards almost completely destroyed by Cromwell. A fragment only remains.

Dalton Castle may be mentioned as a building uniting in some respects the characteristics of the castle and the peel tower. Its early history is unknown.

23. Architecture—(d) Domestic. Manor Houses, Cottages.

In many parts of our district there has been an extensive erection of modern buildings with varied styles of architecture. With these we are not so much concerned, but shall consider more particularly the ancient buildings which present certain features typical of this part of the country.

The more ancient dwellings may be divided into two classes—the manor-houses of the lords of the manor, and the cottages of the peasants.

The more interesting manor-houses were built between the fourteenth and the seventeenth centuries. The characteristic feature of these is the peel tower, rendered necessary for a people liable to raids : it originally constituted the whole dwelling. These towers were modelled on the keeps of the Norman castles. They were rectangular and usually three-storeyed. In the lowest storey were kept stores, the inhabitants occupied the middle one by day and slept in the upper one by night. The roof was used for fighting purposes when raiders had to be repelled. The peel-tower stood in an enclosure called the "barmkyn," surrounded by a wall. Into this barmkyn the cattle were driven when raids were feared. In later times dining halls and other additions were built out from the peel-tower.

Wraysholme Tower (see p. 136), one mile from Kent's

Swarthmoor Hall, Ulverston

Bank, and Broughton Tower at Broughton-in-Furness, are examples of peel-towers, and the tower at Borwick Hall near Carnforth is modified from an earlier peel-tower.

Of other Tudor houses we may mention Swarthmoor
Hall, about one mile from Ulverston, whither George
Fox arrived in 1652, and with which he was connected
for about thirty years; Marsh Grange, on the east side of
the Duddon estuary near the Isle of Dunnerholme, built
to serve as a residence for people employed by the Abbot
of Furness, and the birthplace of Margaret Askew,
afterwards married to George Fox; Kirkby Hall, near
Kirkby Ireleth; Coniston Hall by the shores of Coniston
Lake; and Claughton Hall, on the banks of the Lune
between Caton and Hornby.

It has been seen that, after the Reformation, church
building was checked, and that, as far as ecclesiastical
architecture is concerned, there is little to be said
about the North Lancashire buildings of the Renaissance
period. It is otherwise, however, with the manor-houses,
which continued to be set up during this period. Many
of them were erected in an L-shape, and the peel-tower
disappeared. The classic influence is displayed in many
of the details of these buildings. As examples of the
houses of the Renaissance period we may notice Cark
Hall at the south end of the Cartmel peninsula, a fair
example of an old manor-house, and High Satterhow
near Sawrey, between Windermere and Esthwaite. Much
of the structure of Borwick Hall is also Jacobean.

The view of Graythwaite Old Hall, on the south-west
side of Windermere, will give an idea of the appearance of
the interior of an ancient manor-house.

Many modern buildings retain relics of ancient manor-
houses, as for instance Ashton Hall, south of Lancaster,

the seat of Lord Ashton, and Thurnham Hall, south of this, where the old priest hole still exists.

The earlier cottages of the peasants were usually built of rough stone, often without mortar. They generally

Graythwaite Old Hall

had three rooms on the ground-floor, a sitting-room and kitchen combined, a parlour, and a dairy. Stone steps, sometimes outside the house, led to a loft or sleeping-room. Various modifications naturally occur.

24. Communications—Past and Present. Roads, Railways, Canals.

Study of the map at the beginning of this book will show that the principal lines of communication, whether road, rail, or canal, are chiefly confined to the lower grounds. This is due to two things; firstly the principal towns and villages are situated upon the low grounds, and secondly, unless a very roundabout line is required to avoid high ground, the roads and railways will be carried from place to place along the lowlands. Occasionally, a route must traverse high ground, and then a pass is usually utilised, as for instance that of the Trough of Bowland at a height of about 1000 feet, through which lies the direct route between Lancaster and Clitheroe : this pass has hills from 400 to 500 feet above it on either side.

The pre-Roman inhabitants of North Lancashire, like latter-day barbarians of other countries, no doubt possessed an intricate network of paths connecting hamlet with hamlet. Such paths would be kept open through the undergrowth of the lowlands, and would in some cases extend over higher ridges. Having no definite construction, those subsequently abandoned would tend to disappear, while those which continued in use would show no signs of their formation during pre-Roman days.

The Romans were great road-makers. Portions of their roads yet exist, for they were often carefully constructed of stones, and were therefore more durable than those of much later times.

The principal Roman roads met at the station of Ribchester. The road from the south of England coming through the Roman town of Mancunium (Manchester) here crossed the Ribble and entered North Lancashire. It was continued northward over Longridge Fell, and up the valley of the Hodder, crossing the fells of the Forest

Borwick Hall, Carnforth

of Bowland near Bolton Head at a height of about 1500 feet above sea-level. It then passed northward on the right bank of the Hindburn, crossed the Wenning near Bentham, and entering the valley of the Lune, reached the Roman station at Overborough near Kirkby Lonsdale, after which it entered Westmorland, and went away to Carlisle and Scotland.

From Ribchester three other roads left this main north road. One went a little north of east across the Pennines to the Roman station at Ilkley and so communicated with the important district of the plain of York. This soon crossed the Ribble and so quitted what is now North Lancashire.

In the opposite direction a road was carried to a Roman seaport which stood on the north side of the Ribble estuary at the Neb of the Neeze or Naze Mount, near Freckleton, to the west of Preston.

A third road extended in a direction west of north over the western end of Longridge Fell, and passing along the junction of high and low ground reached Lancaster. From Lancaster another road went up the Lune valley and joined that first described at Overborough. From Overborough a road ran south-eastward into Yorkshire, but this soon quitted Lancashire soil.

An early Roman route seems to have left Lancashire to cross the sands of Morecambe Bay, and pass along the coast of Cartmel and Furness, thence crossing Duddon sands into Cumberland, and so around the Cumbrian coast to Carlisle.

Lastly, a road leaving the Roman station at the head of Windermere went westward over Wrynose Pass near the line of the county boundary, and entering Cumberland ended at the Roman port at Ravenglass.

From the time of the Romans to the early part of the nineteenth century, when McAdam caused a return to the use of stone foundations, the roads of Britain were very bad, and much difficulty was experienced in getting from place to place. The introduction of the method of formation of roads due to McAdam was necessitated by the rapid development of stage-coach travelling about that time.

Of the modern roads we need say little, for those of importance are indicated on the maps on the covers. The great artery of road-traffic extends from Preston to Lancaster, where it branches, one branch going along the low ground to Kendal and northward over the fells, the other up the vale of Lune past Kirkby Lonsdale, to join the former on Shap Fells in Westmorland.

The first Lancashire railway was that between Manchester and Liverpool, opened in 1830, but it was not until 1844–46 that the railway came into that part of the county with which we are concerned. In these years the Lancaster and Carlisle railway was made. Nowadays four important railway companies have lines in the northern part of the county, namely the London and North Western, Midland, Furness, and Lancashire and Yorkshire. The main line of the London and North Western system enters the north of the county at Preston, and running northward through Lancaster, leaves the county north of Carnforth. From Garstang a line owned by a separate company runs *viâ* Pilling to Knott End. From Lancaster branches of the London and North Western railway are carried to Glasson Dock and Morecambe.

The Midland railway enters the county near Wennington and runs down the Lune valley to Lancaster, Morecambe, and Heysham. The Lancaster and Heysham portion is electrified. A joint line of the Midland and Furness companies connects Wennington with Carnforth.

The Furness line starts from Carnforth, and extends along the north coast of Morecambe Bay through Ulver-

Broken Bridge on the Canal, Lancaster

ston and Barrow-in-Furness to Foxfield Junction, where it enters Cumberland. Branch lines go to Lake Side Windermere, and Conishead Priory from Ulverston, to Piel Pier from Barrow, and to Coniston from Foxfield Junction.

The Lancashire and Yorkshire and London and North

Western companies have joint lines between Preston, Blackpool, and Fleetwood ; Preston, Lytham, and St Anne's-on-Sea ; and Preston and Longridge ; and a line also connects Blackpool with Lytham.

An electric tramway runs parallel with the coast between Blackpool and Fleetwood, and similar tramways run in the larger towns.

The Preston and Kendal canal passes through Lancaster and leaves the county near Burton in Westmorland. A branch extends to Glasson Dock. Packet-boats for passengers began to run on this canal in 1820, but have long been discontinued. A short canal connects Ulverston with Morecambe Bay.

25. Administration and Divisions— Ancient and Modern.

It has been seen that Lancashire is of late date as compared with most other English counties.

Before the formation of the county, various divisions existed for administrative purposes. Thus we find in Domesday Book that South Lancashire is treated with Cheshire, and North Lancashire with Yorkshire, and before those shires were created other divisions must have existed.

In the early days of tribes administration was no doubt at first mainly a family affair and later an affair of clans, but as the tracts of land under one ruler increased, it was necessary to make divisions of the tracts, in each of which administration of some of its affairs was local, and accordingly the administration by families and clans was replaced by a territorial one. The early Saxon "shires" of the south, each probably the result of expansion outwards of a definite colony, formed convenient territorial areas for administrative purposes, and under the Normans those areas which had not thus been parcelled out were formed into counties. Hence the existence of the county of Lancashire for similar purposes.

The Saxons had *ealdremen* or governors who appointed deputies called *sheriffs* (shire-reeves; reeve being equivalent to our word steward). The inferior people were partly *ceorls* or freemen, and partly *villeins* who were labourers in the service of particular persons, and not strictly slaves. Upon the establishment of the feudal system by the Normans many of the Saxon laws and customs were retained, as was also the old distinction of classes. Thus there were counts or earls, barons, knights, esquires, free tenants, and villeins.

William the Conqueror bestowed the district upon Roger of Poitou, who ruled it as a count, and divided portions of it among his followers.

The system of local government necessitated the division of the county into minor tracts for administrative purposes. It was separated into *hundreds*, which are supposed to have been so-called because each originally contained a hundred families or freemen. Of these there are six, of which two are in North Lancashire, namely Lonsdale (north and south of the Sands) and Amounderness. A further division was made into *parishes,* each with its own officials, and the parishes were again divided into *townships* or *constablewicks.* As the shire had its sheriff, so the parish had its own special reeve, or presiding official.

The gradual accumulation of numbers of people in restricted areas gave rise to towns, and necessitated special government in the case of these towns apart from that exercised by the constable of an ordinary constablewick with his subordinates.

The powers of government of a corporate town are granted by a Royal Charter, and the first charter possessed by the town of Lancaster is dated 1193. The administrative powers of those responsible for the government of the town were altered by many subsequent charters, and the town is now governed by a corporation of thirty-two members including the Mayor, eight of whom are aldermen.

Other corporate towns of North Lancashire are Barrow-in-Furness, Blackpool, Morecambe, and Preston.

Barrow, being a town of modern growth, only obtained its charter of incorporation in 1867. The corporation, like that of Lancaster, consists of a Mayor, eight aldermen, and twenty-four councillors.

Blackpool received its charter in 1876. There are forty-eight members of council of whom twelve are aldermen.

Morecambe is a town of modern growth. It obtained its charter in 1902. There are twenty-four members of council of whom six are aldermen.

Preston, like Lancaster, was a corporate town in early days, a charter having been granted in 1179. There are thirty-six members of council of whom twelve are aldermen.

When the county was first constituted its voice in the general affairs of the nation was slight. When the great Charter was signed in 1215 one of its provisions was that its articles were to be carried out by knights from each shire chosen in the County Court. Thus the influence of the county in national affairs became more direct. In 1295 the first complete Parliament assembled, and, besides others, two knights were summoned from each shire, two citizens from each city, and two burgesses from each borough. Since then the county has had full share in the government of the nation, and after many changes, the northern part of the county is now represented by six members of Parliament, chosen by the burgesses of the following divisions and boroughs:—the Blackpool division (1 member); Lancaster division (1 member); North Lonsdale division (1 member); borough of Barrow-in-Furness (1 member); borough of Preston (2 members).

Let us turn now to the present government of the county, which has gradually grown out of the old administrative system. The head officer of the county is the Lord Lieutenant, who in some ways represents his Norman predecessor, who under the title of count, earl, or other name, was at the head of affairs. The Lord Lieutenant represents the Crown in the county, and one of his duties is to nominate all Deputy Lieutenants and Justices of the Peace.

The High Sheriff of to-day is to some extent representative of the Norman Sheriff, although his duties are much restricted in comparison with those of the ancient officer, and are chiefly connected with affairs of the law. He is Keeper of the King's Peace within the county, and he attends the judges of the realm when on circuit.

A recently constituted body with purely administrative functions is the County Council. It consists of a Chairman, Vice-Chairman, Aldermen, and elected Councillors. The county is divided into a large number of electoral districts, each of which returns a councillor: from these are chosen the aldermen. Among its functions are the management of county halls and buildings, pauper lunatic asylums, bridges and main roads, the appointment of certain officers such as coroners, the control of parliamentary polling districts and contagious diseases of animals. The County Council are also the local education authorities through an Education Committee. There are also Rural District, Urban, and Parish Councils, for the administration of smaller areas of the county.

Of the public schools in the county the best known are Rossall School, Stoneyhurst College, and Lancaster Grammar School, to which reference is made elsewhere.

For purposes of Justice the northern part of the county has assizes which are held at Lancaster, quarter sessions presided over by a Chairman of Quarter Sessions, and also a number of petty sessions, each having Justices of the Peace, whose duty it is to try and to punish offenders against the law. Civil cases are tried in the County Court, over which a special judge presides.

North Lancashire before 1847 was in the diocese of

Rossall School

Chester. In that year the diocese of Manchester was created, the greater part of North Lancashire being included, but the following parishes in the tract north of the Sands are in the diocese of Carlisle:—Hawkshead, Cartmel, and Colton with their dependent ecclesiastical parishes; Aldingham and Dalton with dependent parishes, Kirkby Ireleth, Pennington, Ulverston, and Urswick.

The dioceses are divided into archdeaconries, these into rural deaneries, and these still further into ecclesiastical parishes, which are not the same as the civil parishes.

26. Roll of Honour.

Foremost among those who have done honour to their county must be placed the great families of whom individuals have through the centuries been prominent for promoting the welfare and directing the affairs of the county, and in many cases rendering important services to the country. It would clearly be impossible in a small work on geography to mention the numerous services of the different individuals of these families, and we must be content to make this general reference. Special mention must however be made of William Cavendish, seventh Duke of Devonshire (1808–1891) to whom the development of Barrow-in-Furness owes so much. He helped to establish the industries of iron-mining and steel production, and was a great breeder of shorthorns, the herd at Holker having had in its time a very wide reputation.

The selection of notable men is difficult ; no two writers would agree as to the list of worthies to be

Archbishop Sandys

placed in the roll of honour. In three cases names have been included of men who were not born in the county, for to the geographer, the accident of birth is not of so great importance as the selection of the county for residence, on account of particular advantages which it may offer, or because the district is favourable for the work to be done.

We will begin with three ecclesiastics in the order of their birth. Edwin Sandys (1516?–1588) was probably born at Hawkshead. He became Archbishop of York in 1575 or 1576. He was one of the translators of the Bishops' Bible, and founded the Grammar School at Hawkshead. A very different man was William Allen (1532–1594), who was born at Rossall, and ultimately became a cardinal. He was a strong supporter of the Roman Catholic religion, and was one of those who urged the King of Spain to invade England. He was also author of many works. Edmund Law (1703–1787), born in the parish of Cartmel, was a writer of many theological works, and became bishop of Carlisle.

Lancaster may be proud of having produced three distinguished men of science. William Whewell (1794–1866), a voluminous writer on many subjects, possessed too varied knowledge to show much originality in any one subject. He was Master of Trinity College, Cambridge, and his reputation rests chiefly on the influence he exerted upon his university, especially as regards its studies. Sir Richard Owen (1804–1892) "was recognised throughout Europe as the first anatomist of his day." He published a vast amount in connection with his subject, much of it being of permanent value. He was also chiefly responsible for the development of the British

Sir Richard Owen

Museum of Natural History, of which he became Director. Sir Edward Frankland (1825–1899) was a distinguished chemist, specially noted for the value of his contributions to organic chemistry.

The same town produced Henry Cort (1740–1800) whose discoveries, including the substitution of coal for charcoal in converting pig-iron into wrought iron, had a most important effect upon the iron-trade.

Sir Richard Arkwright (1732–1792) was born at Preston. He was an inventor of machinery as a substitute for hand-labour in textile manufactures, and his invention, the spinning-jenny, is known to every one. His first machine was set up at Preston, but he afterwards moved into Derbyshire.

Sir John Barrow (1764–1848), born at Dragley Beck near Ulverston, may claim to be the founder of the Royal Geographical Society. He was Secretary to the Admiralty, and is specially noted as a promoter of Arctic exploration. His monument on Hoad Hill near Ulverston is a conspicuous landmark.

One great artist hails from North Lancashire, namely George Romney (1734–1802). He was born at Beckside near Dalton-in-Furness, the son of a builder and cabinet-maker. In addition to being one of the most distinguished of English portrait-painters, he trained others, among whom was James Lonsdale (1777–1839), born in Lancaster.

Of men born elsewhere who had much influence in our area, mention may be made of George Fox, William Wordsworth, and John Ruskin. George Fox (1624–

George Romney

1691), founder of the Society of Friends, was born in Leicestershire, but much of his work was done in North Lancashire. He married Margaret Fell, widow of Thomas Fell of Swarthmoor Hall near Ulverston, and spent many

Sir Richard Arkwright

years there, building and endowing its meeting-house. On several occasions he suffered imprisonment for his zeal, being confined in Lancaster Castle among other places. William Wordsworth (1770–1850) was born in Cumberland, but much of his poetry is descriptive of the hilly district of North Lancashire. Great as a poet, he must also be regarded as one who has claims among the worthies of our county, for the recognition of the Lake District as a place of beauty owes much to his writings. John Ruskin (1819–1900), born in London, author, artist and social reformer, also wrote of the district, which he loved so well that he spent the last years of his life there. He has many references to those Coniston hills, among which nestles Brantwood, and where, in the churchyard of Coniston his remains now lie.

27. THE CHIEF TOWNS AND VILLAGES OF NORTH LANCASHIRE.

(The figures in brackets give the population in 1901. C.B. = County Borough, M.B. = Municipal Borough, U.D. = Urban District. Those not lettered are Civil Parishes. A population of about 1000 is taken as admitting a place into this list, though a few urban districts with a greater population but having no large village are omitted, while a few places of small population but presenting features of interest are inserted. The figures at the end of each section are references to the pages in the text.)

Aldingham (1072). A village on the south-east side of the Furness peninsula, five miles east of Barrow-in-Furness, and four miles south of Ulverston. The church has some Norman arches inside, and its tower is of Decorated and Perpendicular work. The Mote Hill, possibly the site of a Norman settlement, stands by the sea-shore, a short distance to the south of the village. The ruins of Gleaston Castle are about two miles west of Aldingham. (pp. 119, 127, 135, 152.)

Barrow-in-Furness, C.B. (57,586, provisional figures for 1911, 63,775), a county and parliamentary borough and port, is situated at the south-western end of the Furness peninsula opposite to Walney Island, on to which it is now extending. Its rise has been very rapid. At the beginning of last century it was non-existent, and on the early Ordnance maps the smallest type was used for the name "Barrow," indicating that it was then a mere hamlet. In 1837 the population was 100. Its rise was largely due to the seventh Duke of Devonshire. The town occupies a considerable area of the mainland, and extends over Old Barrow Island and part of Walney Island. The regular streets of part of the town, arranged in rectangles, have been already noticed. The chief industries are the manufacture of

Barrow-in-Furness: Devonshire Dock

iron and steel, shipbuilding, gun-mounting, engineering, boiler-making, jute-spinning and weaving, the manufacture of bricks, the importation of timber, paper-making, railway works, cement works, etc.

The principal public buildings are the Town Hall, the Municipal Buildings, Technical School, Barrow Hematite Steel Company's works (started in 1859), and Vickers, Sons and Maxim's works on Walney Island.

The first docks were opened in 1867. The docks all communicate with one another and have about 300 acres of water area. The Devonshire and Buccleugh Docks between Old Barrow Island and the mainland are spanned by a high-level bridge.

Furness Abbey: Sedilia

Cavendish and Ramsden Docks extend south-eastward between Barrow and Piel.

Furness Abbey is a short distance north of the town, and Piel Island with its ruined castle is about four miles away southward.   (pp. 1, 41, 105, 127, 128, 145, 148, 149.)

Blackpool, M.B. (47,348, provisional figures for 1911, 58,376). A municipal borough and watering-place on the coast between the estuaries of the Wyre and Ribble, eighteen miles west-north-west of Preston. It is the head-quarters of a county-court district. Owing to its advantageous position as regards climate and scenery, Blackpool became a watering-place in the early days of resort to the sea—more than one hundred years ago. It has increased greatly of recent years but large as its population is, it forms a very small fraction of the number of

Coniston Village

visitors who annually flock to the town. For the housing and entertaining of these are hotels and lodging-houses, winter-gardens, and nearly three miles of promenade. Pleasure steamers also leave Blackpool for various places. (pp. 44, 59, 61, 96, 97, 146, 148.)

Bolton-le-Sands (902). A village four miles north of Lancaster, close to the shores of Morecambe Bay. The tower of the church is a good example of the Perpendicular style. The Grammar School was founded by Thomas Assheton in 1619. It is now an elementary school. (pp. 117, 119, 128.)

Broughton West (1117), a civil parish containing the town of Broughton-in-Furness, is nine miles north-west of Ulverston. It has a large market-place with an obelisk of the time of George III. Broughton Church has a Norman doorway. The Broughton Tower, a peel tower of uncertain age, is built into a

Coniston : Brantwood

modern edifice. The town is a convenient starting-point from which to view the beautiful scenery of the Duddon valley. (pp. 127, 138.)

Carnforth, U.D. (3040), is a township containing the village of the same name situated 6½ miles north of Lancaster.

Iron-smelting works give employment to many of the inhabitants. Carnforth is an important railway centre and junction of the North Western, Midland, and Furness lines. The limestone of the neighbourhood is tunnelled by several caverns. Borwick Hall, an Elizabethan and Jacobean mansion, lies two miles north-east of Carnforth. Land has been reclaimed at the mouth of the

Fleetwood Docks

Keer, and the work of reclamation is still proceeding. (pp. 43, 53, 65, 99, 138, 145.)

Cartmel. See Lower Allithwaite.

Caton (1181). A township and village on the left bank of the Lune, five miles north-east of Lancaster. Roman antiquities have been found here, including a milestone. The scenery of the Lune valley near Caton is celebrated, and was the subject of a well-known picture by Turner. (pp. 54, 117.)

Colton (1648). A village situated among the Furness Fells, not far from the foot of Coniston Lake. The parish contains several scattered hamlets. Colton Old Hall was formerly the seat of the Sandys family. (p. 152.)

Coniston (1111). A village near the head of Coniston Lake, and at the eastern foot of Coniston Old Man. It is an important tourist centre. Formerly extensive copper mines and some lead mines were worked here, and it is still an important centre for slates, for which large quarries are opened on Coniston Old Man and in Tilberthwaite. Prehistoric hut circles and other relics are scattered over the moor to the south-west of the village. Coniston Hall, one mile from the village by the lake, is a fifteenth century seat of the Flemings, with round chimneys characteristic of the district. John Ruskin lived for many years at Brantwood, on the other side of the lake, and is buried in the churchyard. The Ruskin Museum is in the village. (pp. 58, 65, 67, 139, 145, 159.)

Dalton-in-Furness, U.D. (13,020). A market-town distant nearly five miles south-west of Ulverston, in the Furness iron-ore district. Dalton Castle has been noticed in Chapter 22. The town is mainly modern, owing its rise to the iron-ore industry. Furness Abbey is within easy reach of this place. (pp. 23, 102, 137, 152, 156.)

Fleetwood, U.D. (12,082). A parish, town, port, and watering-place at the mouth of the Wyre on its south side. The docks are extensive and possess a grain elevator. There is an esplanade, and electric trams connect Fleetwood with Blackpool. Knott End, a rising watering-place on the opposite side of the river, is connected with Fleetwood by a ferry. Steamers leave Fleetwood harbour for various places. This place is one of the most important fishery ports in England.

Hawkshead: Flag Street

Rossall School, by the coast, about two miles south-west of Fleetwood, is one of the great public schools in Lancashire. It was opened in 1844. (pp. 44, 46, 55, 101, 105, 146.)

Freckleton (1239). An ecclesiastical parish, township, and village near the estuary of the Ribble, west of Preston. The Neb of Neeze, to the south of the village, is believed to be the site of a Roman port. (p. 143.)

Hawkshead School

Fulwood (5238) is about a mile north of Preston, of which it is a suburb. Extensive barracks are erected here.

Garstang (808). A market-town nearly half-way between Preston and Lancaster on the Wyre, having a triangular market-place with restored market-cross. At Churchtown, south-west of Garstang, is an old church. The ruins of Greenhalgh Castle are close to Garstang. The Garstang Town Trust still manages the public lighting, the Town Hall, and other matters under a charter of Charles II. (pp. 55, 113, 136, 144.)

Grange-over-Sands, U.D. (1993), situated beneath cliffs on the shores of Morecambe Bay at the mouth of the Kent estuary, nine miles east of Ulverston, is supposed to owe its name to having been a grange of Furness Abbey. Owing to its mild climate and sheltered position it has become a favourite watering-place. (p. 42.)

Halton (890). A village 2½ miles east-north-east of Lancaster, on the right bank of the Lune. Remains of pre-Norman

Heysham: "The Stone Coffins"

crosses are preserved in the church and churchyard. A mote-hill, possibly Norman, stands close by the village. (pp. 117, 119.)

Hawkshead (638). A small town near the head of Esthwaite Lake, five miles south-south-west of Ambleside. It possesses picturesque narrow crooked streets. The church is a fifteenth century structure. The Grammar School, now closed, was founded by Archbishop Sandys in 1585. The poet Wordsworth was educated here. Hawkshead Hall, north-west of the town, was a medieval manor-house belonging to Furness Abbey. (pp. 58, 127, 129, 154.)

Heysham, U.D. (3693). A village on the west side of the peninsula between the Lune estuary and Morecambe Bay, on the shores of the bay, about five miles west of Lancaster. The churchyard is a veritable museum of antiquities, especially rich in relics of pre-Norman date. Some of these have been noticed in

Skerton Bridge, Lancaster

previous chapters. The village is also a health resort, and will no doubt ere long be continuous with Morecambe to the north. The place has recently become important on account of the harbour formed in connection with the steamers of the Midland Railway Company which run from here to Belfast and Dublin. (pp. 43, 99, 117, 126, 144.)

Kirkby Ireleth (1477). A village about five miles west-north-west of Ulverston. The church has a Norman doorway.

Kirkby Hall is a sixteenth century manor-house. Large slate quarries in the Silurian rocks occur on the fells above the village. South of Kirkby Ireleth is Askham, with iron-furnaces. (pp. 99, 103, 127, 139, 152.)

Kirkham, U.D. (3693). About eight miles west-north-west of Preston, in the low grounds of the Lancashire plain. A Saxon name derived from the church, which was erected before

Lancaster Grammar School

the time of Roger of Poitou. Kirkham possesses a grammar school.

Lancaster, M.B. (40,329), the ancient capital of the county, is situated on the river Lune where it becomes tidal. Its history is a long and varied one. It was a Roman station and under the Normans belonged to Roger of Poitou, who built the Castle Keep before 1102. Under John of Gaunt the town prospered, but was ruined during the Wars of the Roses. In the Civil War many

battles took place here. The earliest existing town-charter is
dated 1193.

Notwithstanding the antiquity of the town, there are few
ancient buildings in it. Except for the Castle and St Mary's
church there is nothing left of earlier date than the seventeenth
century. Many fine modern buildings have however been
erected, latest and most prominent of which is the Town Hall,
of classic design, presented by Lord Ashton, and opened in 1910.
The Royal Grammar School was founded in 1472, and is the
oldest school in the county. The Storey Institute was founded by
Sir Thomas Storey in commemoration of the jubilee of Queen
Victoria. On a height above the town is the Williamson Park.

Lancaster was formerly a port of some importance, but owing
to the changes in the estuary it has now dwindled. Prominent
among the manufactures of the town are those of oilcloth and
linoleum. (pp. 3, 50, 54, 99, 105, 110, 117, 119, 120, 128, 129,
132—134, 139, 143, 144, 146, 148, 150, 154, 155, 156, 158.)

Longridge, U.D. (3285), situated at the west end of Longridge Fell, is in an interesting hill-district. (p. 146.)

Lower Allithwaite (801) includes Cartmel, a small
market-town 2½ miles west-north-west of Grange. The beautiful
priory church is noticed in Chapter 21. Several other buildings
of interest are in the neighbourhood. (pp. 95, 116, 127, 128,
129, 131.)

Lower Holker (1062), two miles south-west of Cartmel.
Holker Hall, a seat of the Duke of Devonshire, is a modern
building on the east side of the Leven estuary.

Lytham, U.D. (7185). A watering-place on the north
side of the estuary of the Ribble, opposite Southport, with the
usual sea-side attractions. Shrimp-fishing is carried on here.
(pp. 44, 101, 146.)

Marton (1603), two miles south-east of Blackpool.   Marton Mere lies east of the village.   (pp. 59, 61.)

Morecambe, U.D. (11,798), a municipal borough and watering-place four miles north-west of Lancaster, was formerly known as Poulton-le-Sands.   There are the usual attractions for visitors, many of whom come from Yorkshire, with which Morecambe is connected by Midland Railway.   It commands a fine view of the Lake District hills.   Mussels, shrimps, and flat-fish are taken here.   (pp. 43, 88, 99, 144, 148.)

Overton (346). A village four miles south-west of Lancaster, on a mound rising from the alluvial flats towards the mouth of the Lune.   The church has a Norman doorway.   (p. 127.)

Pennington (1510). A village two miles west-south-west of Ulverston.   Ancient earthworks exist here.   (p. 152.)

Pilling (1407), seven miles west-by-north from Garstang, situated on low ground near Morecambe Bay.   There is a gullery at Pilling Moss.   (p. 144.)

Poulton-le-Fylde, U.D. (2223), was formerly the chief market-town of the Fylde district.   Near the old market cross are well-preserved stocks and whipping-post.   The old tithe-barn is in the street of that name.

Preesall with **Hackensall** (1423) is in the Fylde.   The watering-place of Knott End, on the estuary of the Wyre opposite Fleetwood, is in this parish.   Knott End is governed by an urban district council.   Salt-mining is carried on in the parish.   Hackensall Hall was built in 1656.   Parrox Hall, the residence of the Elletsons, lies north-west of Preesall.   (p. 105.)

Preston, C.B. (112,989, provisional figures for 1911, 117,113), is built on fairly high ground on the north side of the river Ribble on the important western highway to the north.   It

is a county, parliamentary, and municipal borough, the chief town
of the hundred of Amounderness, and of Lancashire north of
Ribble, and an important railway centre.

Though the bulk of the buildings are modern, the antiquity
of the town is, among other things, shown by the use of the word
"gate" for some of the principal streets.

A charter was granted to Preston in 1179. The town is
governed by a mayor and 36 councillors, of whom 12 are aldermen. It returns two members to Parliament.

Preston: Art Gallery and Town Hall

In 1307 the town was burnt by Robert Bruce. In the Civil
Wars there was a small contest in 1643, and one of greater
importance in 1648 when the Royalists were defeated by the
Parliamentarians.

In 1715 the Pretender's party obtained possession of the
town, but were compelled to surrender.

Preston is one of the chief centres of the cotton trade, which is the principal industry of the town. The first Preston cotton mill was erected in 1777. Richard Arkwright worked in this town. Another industry is cable-making.

The Port of Preston has modern docks covering 40 acres. A Franciscan Friary stood about half-a-mile west of the market-place, and there was a Benedictine Monastery at Tulketh; the fabric of the churches is modern. The Grammar School is in Cross Street. A fine group of buildings includes the Town Hall,

Stoneyhurst College

a Gothic building by Sir Gilbert Scott, on the south side of the market-place, and the Harris Free Library and Museum to the east of the Town Hall. (pp. 10, 44, 50, 56, 96, 97, 110, 111, 119, 131, 144, 146, 148, 149, 155.)

Ribchester (1237), situated on the north bank of the Ribble, eight miles east-north-east of Preston, lies at the junction of two Roman roads, and the site of a Roman station is on the banks of the Ribble close to the village. Several old halls are in the neighbourhood on either side of the river. At Styd, half-a-mile east of Ribchester, are old almshouses and an interesting little church of the period transitional from Norman to Early English. The Roman Catholic School of Stoneyhurst is about four miles east-north-east of Ribchester. It is one of the best known Roman Catholic public schools in the country; it was originally founded in 1592 and was re-opened in 1794. (pp. 117, 127, 150.)

St Anne's-on-Sea, U.D. (6838). A watering-place on the north side of the estuary of the Ribble west of Lytham. A lighthouse marks the north shore of the Ribble. Shrimp-fishing is carried on here. (pp. 44, 46, 101, 146.)

Seathwaite with **Dunnerdale** (263). A hamlet situated amid the beautiful scenery of the Duddon valley.

Silverdale (582). A prettily situated village on a peninsula, facing the estuary of the Kent. It is a summer resort for visitors. The scenery of the Mountain Limestone district around is interesting. (p. 61.)

Ulverston, U.D. (10,064). A market-town in the Furness peninsula. Iron-smelting is carried on here. There is a Norman doorway in the parish church. The monument to Sir J. Barrow on an adjoining hill is a conspicuous object. Swarthmoor Hall lies about one mile south of the town. (pp. 99, 116, 127, 139, 145, 146, 152, 156, 158.)

Urswick (1186). Three miles south-south-west of Ulverston. There are two villages, Great and Little Urswick, the former built around part of Urswick Tarn. Several antiquities are found in the neighbourhood. The Grammar School was founded in 1579. It is now an elementary school. (pp. 61, 116, 127, 152.)

Warton with **Lindeth** (1482). Warton is a village situated under the eastern cliffs of Warton Crag at the head of Morecambe

Bay. The Perpendicular church has the shield of the Washington family built into the tower. The first President of the United States was descended from this family, and it is said that the American banner, "The Stars and Stripes," was derived from the Washington Arms. On Warton Crag are many antiquities,

Yealand Village

including the promontory fortress on the summit having its ramparts on the north.

Yealand Conyers (267) and **Yealand Redmayne** (191) lie on the east side of a limestone hill two to three miles north of Carnforth. The villages are amongst the prettiest in the north of England.

Note. The figures in the following diagrams refer to Lancashire as a whole.

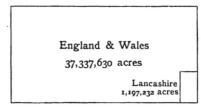

Fig. 1. The area of Lancashire compared with that of England and Wales in 1911

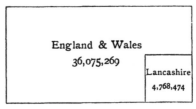

Fig. 2. The Population of Lancashire compared with that of England and Wales in 1911

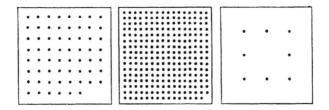

England and Wales 618   Lancashire 2550   Westmorland 80

Fig. 3. Comparative Density of Population to the square mile in 1911

(Each dot represents 10 persons)

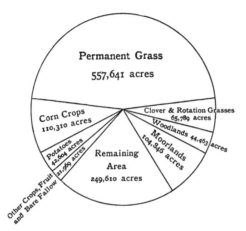

Fig. 4. Proportion of Permanent Grass to other areas
in Lancashire in 1910

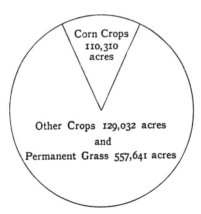

Fig. 5. Proportionate area under Corn Crops compared
with that of other Crops in 1910

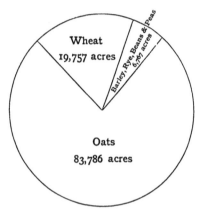

Fig. 6. Proportionate areas of Cultivated Land in
Lancashire in 1910

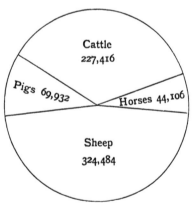

Fig. 7. Proportionate numbers of Live Stock in
Lancashire in 1910

Milton Keynes UK
Ingram Content Group UK Ltd.
UKHW032321161024
449665UK00001B/9